Praise for **A Planet of 3 Billion**

Like Rachel Carson's Silent Spring, more than a half century ago, this book should serve as a wake-up call to a generation that is more in tune than ever with our planet's pulse. As Tucker proclaims, our planet has a people problem. And, as humanity and nature struggle to coexist sustainably, it is time for all of us to focus all our efforts on bending the human population curve downward. Tucker's formula begins with the empowerment of women, worldwide - but his cookbook for global citizens helps guide our collective actions to a resilient future. A Planet of 3 Billion is essential reading for anyone who cares about the fate of our planet and our species.

Dr. Jane Goodall, DBE
Founder - the Jane Goodall Institute
& UN Messenger of Peace

It is easy to marvel at the sheer enormity and growth of our world, but it is irresponsible to do so without reckoning with its finiteness. In A Planet of 3 Billion, Dr. Christopher Tucker adeptly argues how we can (and we should) shrink the global population from nearing 9 billion to a more manageable 3 billion—and in so doing repay our debt to the planet, a burden it no longer can bear. Timely and instructive, Dr. Tucker's work explores the nuanced and often overlooked relationship between geography, women's empowerment, and leadership— an intersection which could have tremendous impacts on our future.

General Stanley A. McChrystal, USArmy (Ret.)

Tucker has written an exceptionally broad-ranging, thought-provoking examination of the relationship between expanding human numbers, socio-economic innovations, and ecological degradation. His effort to explore the historical and geographical roots of our current environmental predicament challenges us to think anew about the future of our planet.

Professor Alexander B. Murphy
Rippey Chair of Liberal Arts & Sciences, University of Oregon
Fmr. President, American Association of Geographers

A sweeping analysis of extraction, combustion and pollution spelling an unsurvivable future for our species unless we adopt dramatic changes in ecosystem protection and practices. It is irresponsible not to read this book.

Vint Cerf
Internet Pioneer

Christopher Tucker asks and answers the central question of the 21st century, that being, how do we build a habitable planet? He answers that we build it in the way humans have always advanced before we lost our way. We had a plan and we adjusted to an environment. Now, we are determining the environment. We have become the designer species, but without a design. An invasive species with a brain, but no method to plan. Tucker outlines a thoughtful action plan for long term adaptation and success and does so with the complete set of tools and ideas and theories that will allow for massive human success going forward. This book breaks down the walls of our conceptual prison and offers a path to design freedom for the future. A path we sorely need to find.

Michael M. Crow
President and ASU Foundation Leadership Chair
and Professor of Science & Technology Policy
Arizona State University

To Ann, my North Star.

A PLANET OF
3 BILLION

**Mapping Humanity's Long History of Ecological Destruction
and Finding Our Way to a Resilient Future**

A Global Citizen's Guide to Saving the Planet

CHRISTOPHER TUCKER

ATLAS OBSERVATORY PRESS
WASHINGTON, DC

Published in the United States by Atlas Observatory Press.

Atlas Observatory Press books may be purchased in quantity for educational, business, or promotional use. For information please send an e-mail sales@atlasobservatorypress.com.

Set in Palatino type by Reynaldo Dizon, Jr.

Printed in the USA, by IngramSpark.

Library of Congress Cataloging-in-Publication Data

Name: Tucker, Christopher Kevin.

Title: A Planet of 3 Billion / Christopher K. Tucker

Description: First edition. | Washington, DC: Atlas Observatory Press, 2019

Includes bibliographical references (p.) and index.

ISBN 978-0-578-51530-4
LCCN: 2019909231

Library of Congress Catalog-in-Publication data available.

www.atlasobservatorypress.com

10 9 8 7 6 5 4 3 2 1

First Edition

Book design by Reynaldo Dizon, Jr.

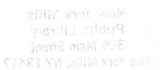

Table of Contents

Foreword

How many times have we been told that we are at an "inflection point"? Any number of books have helped to shed light on impending historical shifts. These books promise that if we view the world through a particular lens, are forward thinking, and poised for action, then we will be able to be a vital part of the future. This is not that book.

Still other books highlight the emerging technological innovations that, if properly harnessed, will help us avert disaster. As they relate to human population dynamics, these books might be characterized as the works of "techno-optimists" who promise us a future free from the miseries first anticipated by Thomas Malthus. This is not that book either. This book instead asks the reader to grapple with one of the most gnarly, wicked problems that humanity will ever deal with and to help us collectively navigate our way to a sustainable future. I request that you ask yourself the question, "How many people can Earth support?" and then go along on an intellectual journey that leads you to some uncomfortable realities about our planet's actual ecological carrying capacity. Given the finite geography of our planet, how many modern humans can it actually support without incurring ecological debt that undermines its ability to support our species and the long-term viability of the wildernesses from which we evolved? From the title of this book, *A Planet of 3 Billion*, you can see where this is going.

This book draws on a rigorous accounting of humanity's rich past to place today's challenges in a historical context that makes them analytically tractable. It wrestles with our complex present to identify the vital trends that will reshape our planet in the coming decades—trends for which we have already laid the foundation. And it investigates the uncertain future that we will have to navigate together as local communities within a global society if we are to evade the perils and pitfalls that we, as a species, have put in play. Part of the uncertainty is due to exogenous factors that simply cannot be anticipated. But the most important factors driving this uncertainty are the choices, both individual and collective, that we all can make to improve the lot of our fellow man and woman, while investing in a more sustainable planet.

In the end, this book focuses on choices that we face, as well as the difficult, deliberate work that will be required if we are to prevent some very bad things from happening. Some will ask, "What bad things?" seemingly oblivious to what is going on. Others will ask, "Which bad things?" typically with some amorphous grasp of the notion that the sea levels may rise if something is not done soon. Yet few have sensed the dangers that lie ahead. This, I would argue, is because no dominant narrative has organized and articulated these dangers in a way that is accessible to the layperson.

This book starts with a rigorous accounting of our past because at this important moment, humanity does not share a common understanding of the challenges that we face. This is because we do not share a common sense of where we are or how we got here. The end of the 20th century and the dawn of the 21st century delivered us a global society composed of many fractured publics, each with very different understandings of where we are in our world's history, and no collective grasp of the common challenges that we face. First and foremost, this book attempts to help everyone get on a common page so that we can have a real discussion about the fate of our planet and our species.

Beyond this message and its urgency is a deeper worldview that I believe is even more fundamental to how we should be thinking about our planet. I believe that all of us, from the everyday private citizen to our vaunted world leaders, must learn to think geographically about the challenges that face our world. After all, everything that has happened on Planet Earth has, by definition, happened in space and time—that is, geographically, as history has unfolded. And to understand the peril that we face, one must first map the long history of ecological destruction that the human species has wrought on our planet, if we are to understand the foundations upon which we can build a resilient future.

As humanity's numbers surpass 7.5 billion (circa 2017), on their way to 9 billion, 11 billion, or more, how should we think about the challenge, and what possibly could be done, if my estimate of Earth's carrying capacity is correct? This question will require as much of your intellectual energy as did the first half of the book, since I am asking you to consider my proposals and think creatively about how you might translate them into action.

While each of us experiences our own truth, during our own historical moment, and in the places we have inhabited, thinking geographically allows us to derive larger truths by organizing and sharing our collective experiences and observations across space and time. By doing this, we can see things that alone we are incapable of seeing. Alone, we cannot see over the horizon. Alone, we cannot know how distant landscapes have been shaped by distant peoples. Together, however, we can understand the world around us—its rich past, complex present, and even its uncertain future—if together, we think geographically.

Geography is a discipline that has seen more than its fair share of derogation over the past half-century. Yet the rise of geospatial technologies and data, including geographic information systems and the complex of remote-sensing technologies, has revolutionized how every discipline and profession has been able to measure and understand our world. This has resulted in the so-called "spatial turn" in the social sciences that has recognized the importance of geography to their analysis and worldview. Although history has long been considered a counterpart to geography, the rigor of time came to geography only with the emergence of these geospatial technologies. And true spatio-temporal technologies and data analytics, in which geographical observations are rigorously anchored in time, arrived in robust form only over the last decade or so. It is this aspect of the geospatial revolution that has made this book possible in the first place.

As such, some have told me that it is silly to write an old-school paper book on this topic, given the dependency that my argument has on enormous volumes of digital spatio-temporal data. Despite their admonitions, I chose to take on the effort of writing this book for one reason: words still have power. Yet maps, particularly when animated over time, have a power all their own. Since, in the end, I am asking everyone on Earth to take on a huge challenge, and to marshal their time, commitment, resources, and ingenuity to help our planet and our species, I believe that it is incumbent upon me to show my data, in a visually compelling way. So, in order to understand the global changes that have brought us to where we are today, over so many spatial and temporal scales, we have also taken care to publish our underlying data on the web at www.planet3billion.com. On this public website loaded with openly licensed data, everyone from children and teachers to scientists and policymakers, and all the everyday citizens in between, will be able to investigate the data that I use to support my argument.

Yet this book is meant to be more than a wake-up call with a detailed marshaling of evidence. I am asking you to come along on a journey where you will learn things about our planet that you likely did not know, or at least had not had the opportunity to integrate in your mind's eye over space and time. Perhaps you will come to the same conclusions as I have about Earth's carrying capacity. Or perhaps you will want to challenge me with your own data and your own geographical accounting of our planet's past, its current disposition, and its future prospects. Indeed, I welcome that.

In either case, I hope that you will find value in the larger debates that this book wades into and attempts to reframe. After all, at some point, and sooner rather than later, the world's population will have to plateau and decrease or else face dire consequences. Will that be when we reach 9 billion? 11 billion? 13 billion? Regardless, at this point, our economy will, by definition, contract and we will experience economic de-growth—a concept for which our economic system is fundamentally unprepared. As populations decrease, new geostrategic challenges and opportunities will unfurl in front of us. This complex situation can only be navigated in a world populated by global citizens who can think critically about the collective action required for us as a species to sustainably coexist with the planet that gives us life. Being a global citizen requires that you think geographically. Without a shared geographical understanding of our planet, our species, and the civilizations we have created, we will soon find ourselves unable to deal with the unfortunate consequences of ignoring certain realities about our planet.

CHAPTER ONE

There Is No Invisible Hand

I am an unrepentant capitalist.[1] The transition from burning wood to burning fossil fuels that capitalism and industrialization brought us was a revolution that very well may have saved humanity, our forest ecosystems, and our planet from impending doom. The proto-capitalism of the First Industrial Revolution made it possible for living standards to rise in ways that just a century earlier would have been unimaginable. Some are inclined to quote Adam Smith, the earliest and most astute observer of the political-economy dynamics at the heart of the Industrial Revolution, and attribute all of this progress to some invisible hand of capitalism. That invisible hand does not exist.

---

[1] As you read on, you will also see that I have many criticisms of how the practice of capitalism has evolved over the past few centuries. Still, the rapid progress that humanity has enjoyed over this time period can only be understood as the product of profit-motivated action by innovators and their financiers organized around firms, both big and small. Nevertheless, the ideology that has been advanced by some capitalists, and their enablers in the field of economics, must be set aside. Economic advancement can only truly be understood within the context of a rich set of supporting institutions including governments, universities, industry associations, labor organizations, and non-profit organizations of many kinds. This coevolution of technologies, industries, and supporting institutions has unleashed a mode of industrial innovation on our planet that has provided unprecedented improvements in humans' quality of life, but has brought with it an alarming level of ecological destruction. This ecological destruction is not inherent in capitalism, but rather in the way in which modern societies have consciously or unconsciously chosen to shape economic actions without sufficient commitment to the ecological consequences of those actions.

Complex political economy, a body of thought that emerged from the moral philosophy of the Scottish Enlightenment, asked tough questions about how to improve humans' lot in life. It was a debate, not an ideology. Industrialization and theories of political economy were part and parcel of each other, as recognized by Adam Smith in *An Inquiry into the Nature and Causes of the Wealth of Nations* (1776), Robert Malthus in *An Essay on the Principle of Population* (1798), and David Ricardo in *Principles of Political Economy and Taxation* (1817). Specialization and division of labor. Profit and capital accumulation. Population and economic growth. Comparative advantage and trade. The proper role of government in shaping economic growth and social welfare. These issues swirled in a vigorous debate anchored in the time and context of the First Industrial Revolution.

Although technological innovation as a source of economic growth was not made explicit in their language, these thinkers firmly planted the seeds of these concepts in the fertile ground of Western thought. It is the debate among these men that has permanently shaped modern society's faith in technology. Adam Smith shined a light on the power of economic specialization for generating wealthy, growing societies and economies. Malthus sounded an alarm about his concern that capitalism would ultimately be unable to feed the growing population that it induced. Ricardo argued that comparative advantage and trade would bring this imbalance between population and resources into equilibrium. And neo-Ricardians concluded that the emerging technological innovations at the heart of this process of continuous specialization and comparative advantage would always enable our economies to care for populations, no matter how fast they grew or what dilemmas they might face.

I am not only an unrepentant capitalist. I am also a technological optimist. Technological innovation has improved humanity's lot in powerful and amazing ways since the days of Smith, Malthus, and Ricardo. At first blush, it would seem from all this technological innovation that Smith's "invisible hand" of market exchange has successfully channeled self-interest toward socially desirable ends. It is certainly more than fair to say that Malthus's dire forecast has been thwarted innumerable times by the wonders of technological innovation, as advanced by the engine of capitalist economic growth, albeit supported by important public and private institutions.

But to say that some invisible hand at play within our political economy has led us inexorably to a sound and sustainable world would be to ignore the heavy ecological footprint that humanity has managed to impress upon our planet. The field of economics has failed to grasp the inherent necessity of the ecosystem goods and services that our planet generates to sustain our species and others—and has utterly failed to account for the ways in which humanity's industrialization of the global landscape has systematically excised enormous portions of nature that generate these ecosystem goods and services on which we depend. Mainstream economics' inability to account for the ecological debt that humans have accrued over time has led humanity—whether average citizens, leaders of industry, or policymakers and political leaders—down the garden path. Exalting a reasonable faith in market mechanisms to a form of unquestionable religion has led us to ignore the ecological devastation that humanity's economic action has wrought.

The progress offered to humanity by modern technological innovation has decreased mortality and increased longevity, driving population growth and consumption patterns. Where Malthus was wrong in thinking that the capitalist economy cannot keep up with the care and feeding of this enormous growing population, he and his colleagues were silent on the other side of this equation—humanity's ecological footprint and its impact on Earth's carrying capacity. This is excusable, in a way, when one reflects on the fact that they published their thoughts in an age when the world had fewer than a billion people, and much of its land was unexplored. However, it is time to move past some notion that the invisible hand will guide our global society through the shoals that we are about to encounter.

When it comes to the future of our planet and the fate of humanity, there is no invisible hand.

CHAPTER TWO

A Finite World with Infinite Possibilities

When I was a student at Columbia University, I had the good fortune to work under Michael M. Crow when he first conceived of the Earth Institute, as the Vice Provost for Research. Watching him breathe life into the Earth Institute concept was a huge learning moment for me. How were we to transcend the intellectual silos of the traditional academic departments and create an interdisciplinary, even transdisciplinary, university-wide unit focused on the big questions about our planet? Front-row seats to that adventure taught me more than I ever could have hoped for, about both understanding our planet's complexity and the complex process of organizing societal initiative to build a sustainable planet. Its inaugural lecture series, held in the historic Low Memorial Library Rotunda in 1995, left a lasting impression.

Simon Schama, the Old Dominion Foundation Professor, lectured on his book *Landscape and Memory*, offering a wide-ranging tour of the human experience and the "psychic claims that humans have made on nature," a talk that reshaped how I viewed the history of our world and the role nature played in it. Wallace "Wally" Broecker, the Newberry Professor of Geology, discussed the 4.5-billion-year evolution of Planet Earth and both historical and present-day mechanisms changing Earth's atmosphere, a talk that put the modern industrial pace of greenhouse gas output in stark perspective. Paul Olsen, the Storke Memorial Professor of Geological

Sciences, reflected on the historical interactions that life has had with the environment, going back 3.8 billion years, to include the various mass extinctions that he speculated may have been due to the planet reaching its carrying capacity at various times.

The theme of the day, however, came from Joel Cohen, Professor of Populations at Columbia and Rockefeller universities, and his new book *How Many People Can the Earth Support?* Cohen's was a great lecture. So don't get me wrong. But I always found it strangely mechanical and soulless. It seemed like his mathematical solutions for determining how many people Earth could support—its carrying capacity—under different scenarios offered a strange moral equivalence between a world where 3 billion people live with a light ecological footprint and a world where 20 billion people live, stacked like cordwood, optimistically yet desperately seeking to construct a technological complex that allows everyone to live to their hearts' content while deftly avoiding the ecological consequences. While Cohen's argument focused on the inputs required to support populations of various sizes—inputs such as water, food, and energy—it seemed to be void of any real discussion of the ecological footprint that humanity projects on a fragile, finite planet under different "production functions"—to borrow a clumsy term from economists for combinations of inputs that produce the maximum level of economic output.

Cohen's book and its central question have haunted me for nearly 25 years. What about his argument left me wanting? What was missing from his argument that might have given me the moral dimension for which I felt the need? Ultimately, it turned out to be geography.

Geography, you say? Well, ours is a finite world, after all. Earth is a world (currently) comprising seven continents and five oceans (depending on how you count them), each with distinct terrain housing unique ecosystems, over and through which water flows in singular ways. Over the past couple hundred thousand years, these land masses and bodies of water have been host to a passion play of sorts, as humanity (or, perhaps, multiple humanities?) has pursued its destiny, often at the expense of the natural world from which we have sought to set ourselves apart.

Through a geographical lens, the mathematical models of Cohen's worldview were suddenly laid bare. It all made sense to me. One could actually make reasonable assessments about the carrying capacity of Earth by looking at how humanity has spread across the globe, how it has grown in size and concentration, and how it has, over time, affected various ecosystems, watersheds, the abutting oceans, and the atmosphere that sustains life.

Geography and time are the keys to unlocking Cohen's question. Spatio-temporal data, and fairly recent techniques for analyzing them, offer us the opportunity to make a net assessment of humanity's impacts on Earth's complex, interdependent parts. By taking advantage of this opportunity, we can not only provide meaningful answers to this central question, we can also help shed light on the strategic questions that must be answered if we are to successfully navigate our way to a new, sustainable population plateau. This would be a population plateau that has a specific geographic distribution, with specific ecological, cultural, economic, and geostrategic consequences. And this plateau would have a specific number.

The premise of this book is that Earth's carrying capacity for humanity is actually only about 3 billion—a number that we hit and promptly surpassed in the mid-20th century. As a consequence, I also conclude that humanity and the planet that supports it are currently living on borrowed time. In effect, humanity has been on a century-long binge, featuring exponential population growth, continuous growth in industrial output and individual consumption, and the ecological devastation that goes with it.

"Devastation?" Yes, devastation. Perhaps this ecological devastation has occurred beyond the horizon of your direct observation. Or perhaps the form of devastation has been invisible, manifesting in chemical reactions that are not visible to human ocular vision. Or perhaps the devastation is not visible to you because it is simply a continuation of the "new normal" that you were born into, continuing the landscapes of your childhood of which you grew fond. All of us have demonstrated the capacity to fall in love with landscapes of whose origins or fundamental transmogrification we have little understanding.[2]

But without commonly known facts arrayed geographically, made viewable through a geographic lens as they have unfolded over time, it is hard if not impossible to see where we are in this passion play and what the consequences are for the future. Luckily, over the past couple decades, the rise of the World Wide Web and the revolution in geospatial technologies and data have changed the lens through which we see our world. And by assembling basic facts geographically and temporally (as we have done at www.planet3billion.com and www.mapstory.org), we may just be able to chart a course by which we can navigate our way to a sustainable future.

As you read this book, you will see that at this stage, I strongly believe that this issue of ecological devastation can still be framed in terms of a debt, rather than a lost cause. But it is a debt that must be paid down promptly and aggressively, or else the loan shark from whom we collectively borrowed may very well ask us to face our debts in a way that we may find brutally painful to many, and lethal to some. I believe that this debt must be paid down promptly by getting our population

---

[2] This has been called the "Shifting Baseline Syndrome"—a phrase coined by fisheries scientist Daniel Pauly in 1995, when describing how each new generation of fisherman uses the ecological baseline from their childhood to evaluate the extent of environmental degradation, failing to appreciate the degradation that occurred under previous generations' watches. In the absence of rigorous historical information or experience with historical conditions, members of each new generation accept the situation in which they were raised as being normal. We lack long timelines of longitudinal data for most ecological phenomena, across both space and time. This leads to people's accepted thresholds for environmental conditions being continually lowered.

back down to 3 billion soon after 2100, and that we must also aggressively clean up after our century-long binge, or else extremely bad things will happen that will lead to ecological and societal collapse on a scale that most simply cannot fathom.

To most, this will sound ridiculous, particularly given the various population projections proffered by the United Nations and other groups. Optimistic assessments currently have population growth leveling off at 9 billion by 2050. Pessimistic assessments have us blowing right through 13 billion by 2100. And these assessments are separate from prognostications about how we might lighten the ecological footprint of people in both the developed and the developing worlds.

Yet I believe that if the issues are framed correctly for a global audience, allowing us all to collectively occupy the right frame of mind, it is entirely plausible that we could get to a new plateau of 3 billion soon after the year 2100—averting a cataclysm of epic proportions. This would require the widespread adoption of particular interventions at the micro, meso, and macro scales. Yes, "think globally, act locally"—but also act regionally and globally, too. It would also demand that we commit ourselves to restoring essential ecosystems so that they can get back to providing our planet with the ecosystem goods and services on which our planet and our species depend.

CHAPTER THREE

Earth's People Problem

For most of my life, whenever a discussion that involved the term "overpopulation" has erupted, the smartest person in the room tends to break out his or her argument about why "Malthus was wrong." Such people are technological optimists, like myself. They will marshal all manner of examples of technological innovations that have always emerged in time to show society a righteous path forward just as we have been faced with navigating various dilemmas. If only Malthus were even relevant to the current discussion, that would be great. Alas, he is not. Our planet actually suffers from a very different "people problem" than that which Malthus popularized. But the very fact that Malthus occupies the popular mind when we think of such issues means we should begin our discussion about Planet Earth's people problem with him.

Thomas Robert Malthus was a cleric and scholar who specialized in issues of political economy and demography. He came to prominence for his 1798 book An Essay on the Principle of Population. His was a simple argument. He argued that population grows geometrically and that the food supply could only grow arithmetically. It is important to note that at the time, protein and calories were finite and often difficult to access, leading to literal starvation—a reality that is harder to imagine today.

Although many modern interpretations of Malthus characterize him as saying population would outstrip the available food production capacity, his actual argument was that whenever there were productivity increases in agricultural production, leading to more food supply, population would quickly grow to eliminate the abundance. The term "Malthusian" has come to characterize a belief that population tends to increase at a faster rate than the food supplies on which it might subsist. So Malthusians believe that unless a population's behavior is governed by some form of moral restraint or unless it is checked by the occasional disaster, the population will be doomed to a fate of poverty and misery. Let's just say that the optimists among us hate to lend any credence to Malthus or anything that might smack of a Malthusian worldview.

Many strains of Malthusianism have evolved over the years, such as "neo-Malthusianism," which holds that since population growth is exponential, it will outstrip the available food resources if not for artificial measures of birth control. Obviously Malthus predated the dawn of effective contraceptives and other forms of birth control. He was also a product of his times, wherein the fruits of the Industrial Revolution were leading to a steady rate of population growth, leading to concerns over how to keep this population fed, particularly on the finite agricultural landscape of the British Isles. But think of it: For humanity's previous 200,000 years of existence, the human population barely grew at all and even fell from time to time. This growth rate averaged less than $1/10$ of 1% annually. So, the population explosion observed during the dawn of the Industrial Revolution required new concepts and led to significant intellectual hand-wringing.

Ultimately, the demographic implications of Malthus's political-economy worldview were cast aside by the arguments of fellow political economist David Ricardo. Ricardo, who became a Member of Parliament after achieving wild success in finance, was a public intellectual of sorts and a member of Malthus's Political Economy Club. Ricardo argued that Malthus failed to appreciate the benefits of economic specialization and the accompanying power of trade. If only our economies each specialized in the production of the goods for which they have some resource advantage and then traded their bounty, it would be more than possible to feed a population that is growing geometrically. Neo-Ricardian arguments that focus on the technological innovations attendant to this process of economic specialization ultimately won the day in most debates over the impending doom envisioned by those who embrace the term "overpopulation." How could there be *too much* population if everyone is increasingly enjoying the bounty of economic specialization, innovation, and the resulting benefits of trade?

For the past two centuries, this worldview has largely guided the narrative advanced by the academic discipline of economics, as well as capitalists extolling the virtues of their economic model. After all, the Industrial Revolution did unleash a wondrous march of progress that has steadily raised standards of living across the globe, albeit extremely unevenly, over the past two centuries. Without the emergence of this economic machine, it would have been inconceivable that Earth's population could have grown as it has. In a way, the Industrial Revolution short-circuited the laws of nature that kept humanity's total population well below 1 billion for its first 200,000 years.

But both Malthus and Ricardo, and their subsequent acolytes, miss the point. It is not how many people Earth can *feed*. It is how many people Earth can *support*. One must remember that the term "support" means that the planet must be able to carry the weight of a population, including both its need for nourishment and its proclivity to burden, if not destroy, the ecosystems on which it fundamentally relies. It is easy to simply focus on the components that nourish humanity in the narrow sense and to aggressively concentrate on technological solutions for keeping up with an ever-growing population. It is much harder, however, to deal with the complex set of interactions that lead to displacement, conflict, disease, disaster, blight, scourge, denuding, spoilage, and collapse. It is not just the challenge of feeding humans, but of preserving an Earth resilient enough to sustain human life in the face of the inevitable adversities that will befall us. It is about a human society whose size is balanced by a planetary ecological carrying capacity that demonstrates a certain resilience.

Malthus and Ricardo not only preceded concepts of contraception. They also preceded concepts of ecology and humanity's dependence on Earth's many and sometimes distant ecosystems—the ecosystems that modern modes of human settlement necessarily disrupt, harm, or even annihilate. Although the invention of "Nature" as a concept was under way by one of their contemporaries, Alexander von Humboldt, it would be a while before his concepts would become available to them. Humboldt occupied a very different set of academic disciplines and professional circles, and it took decades for his concepts, such as biospheric dependencies between forests, the atmosphere, and the oceans, to enter the broader scientific consciousness.[3]

---

[3] It is hard for any of us today to imagine a world without the concept of nature. Yet it took the life's work of Alexander von Humboldt to invent the notion, as Andrea Wulf eloquently argues in her book *The Invention of Nature: Alexander von Humboldt's New World*. Born in a world at the end of the Age of Exploration but still in the throes of colonialism, this Prussian polymath, geographer, explorer, and naturalist awoke to the natural world as being more than a thing to be conquered and commanded by humanity. As a scientist studying the dynamics driving the natural and biological world, he led humanity to consider how its own endeavors affected nature itself. This work led him to become, by some estimates, the 19th century's most famous scientist, mourned around the world upon his death in 1859 and celebrated around the world on the centenary of his birth in 1869, including the unveiling of a bust in New York City's Central Park. Humboldt's writings inspired many scientists and naturalists, including Charles Darwin, Henry David Thoreau, John Muir, George Perkins Marsh, and Ernst Haeckel. Although more natural places and species are named after Humboldt than after any other person, his personal fame faded in the 20th century.

In the end, we need to rebalance our discussions about Earth's people problem. We must go beyond Smith, Malthus, and Ricardo's considerations of human economic productivity and its contribution to our feeding, sheltering, and general wellbeing. We must place such considerations in the context of the natural Earth on which our species evolved and which fundamentally sustains life. Ours is a finite planet, and humanity's productive endeavors have long consumed—proportionately, if not absolutely—the planet's finite natural resources and exhausted the planet's finite ecosystems while spewing all manner of wastes across vast geographies. This has only been magnified as human population growth has continued unchecked in recent centuries. We must think beyond economics if we are to properly understand how many humans Earth can actually support over the long run. We must look to geography, and to the history of our world through a geographic lens.

Understanding Humanity Within Earth's History

When it comes to understanding the state of our world, so many charts and graphs and statistics are proffered to us by special interests with agendas, scholars with preceding intellectual commitments, governmental and intergovernmental agencies with missions to execute, and prolific morons who simply spout nonsense at the top of their lungs. Amid the cacophony, it is hard for any of us to make sense of the state of things.

All current human inhabitants of Earth face the challenge of wondering which act of humanity's passion play they have walked in on. What happened before our current conception of "normal" came to be? Are we at a historic high or a historic low? Are we experiencing an uptrend or a downtrend? Are we at a historic inflection point that bodes well for humanity? Or are we entering a time of tumult that will demand much of us? What would we even measure to make those determinations? And what is the time frame over which these measurements would need to be assessed for them to have any relevance or meaning?

To understand the state of things today and to understand where we are going, we must first back up the clock and understand where we are within the grand sweep of history. This is the "temporal" part of the spatio-temporal data that should inform our worldview. We have all seen charts and graphs and statistics that show the march of time along some particular axes. It is not until we place these longitudinal statistics within a geographic frame that we can understand their consequences for our finite Earth.

Who cares that the global swath of deforestation has recently hit some alarming peak if we do not understand the specific geographies affected and the ecological, economic, social, and geostrategic implications of that deforestation? This demands that we understand such dependencies geographically. Yet most debates about the major trends shaping our planet seem to be both ahistorical and ageographical.

It is the geography of things as they have evolved over time that provides us the insights that we need in order to determine which act of the passion play we are now in—a play in which each of us has only recently been cast.

So, let's start at the beginning.

But let's first agree to do it through a particular lens. Some are inclined to ignore the same bodies of scientific knowledge that make their computers, cell phones, and televisions work. Or to deny the same science that has enabled modern agriculture and our ability to stock the shelves of our grocery stories. Or to openly mock the scientific insights that enable the miracles of modern medicine, that they will demand when they fall ill. In this book we are not so inclined. Although there are and always will be legitimate scientific debates at the continuously advancing frontiers of science, this book will stipulate the basics of physics, chemistry, geology, biology, and geography. This, of course, is required if we are to understand Earth's history, going back long before the dawn of humanity. Let us start with geological time.

Beginning with Geological Time

On the basis of radiocarbon dating, Earth is commonly dated at 4.5 billion years of age. The first observations of life in the geological record date back 3.8 billion years, with the evolution of single-celled organisms. Earth's surface was covered by an enormous ocean, with the first supercontinent (known as Vaalbara) emerging from the oceans 3.6 billion years ago (3.6 million Ma, in geological parlance[4]). The microscopic life that was abundant in the oceans gave way to more complex life, until the Cambrian explosion (541 to 485 Ma) at the dawn of the Paleozoic era (541 to 252 Ma), which, among other things, brought the first wave of land vegetation. In the hundred million years before the supercontinent Pangaea formed at about 300 Ma, not only did modern plant biology evolve. Basic tetrapod biology (amphibians, reptiles, mammals) including the amniotic egg also evolved, making animal life on land possible.

Thus began a long, tumultuous history of living animals co-evolving with terrestrial, freshwater, and ocean ecosystems and the atmospheres that they helped to shape. The first dinosaurs evolved about 240 Ma and served as the keystone animals, as plate tectonics continued to tear Pangaea apart and led to the slow formation of the continents that we know today. When dinosaurs went extinct about 65 Ma (other than the remaining species of what we now call "birds"), the world's continents would have been recognizable in their present general shape, if not in their flora, fauna, and climate.

This tectonic process isolated species to evolve in unique environments, and from time to time, the collision of landmasses unleashed biological exchanges that produced major extinctions. The physical (geographical) separations caused by hundreds of millions of

---

[4] The historical range of this book requires the reader to be familiar with two dating notation conventions. For specific dates in geological time we use Ma (mega-annum), for a million years. For specific dates from prehistoric times forward, we use BCE and CE, which are equivalent to BC and AD.

years of plate tectonics, sea-level rise and fall, glaciation and glacial retreat, and land subsidence created countless unique islands of genetic diversity which led to the evolution of a blinding array of plants and animals. To be more precise about it, these natural processes led to the evolution of a blinding array of life across what we now recognize as the six kingdoms of life: plants, animals, protists, fungi, archaebacteria, and eubacteria.

Did exogenous forces such as volcanoes and meteors play a role from time to time? Indeed, in important ways. But the geological record—if we are to believe the basics of physics, chemistry, geology, biology, and geography—paints a shockingly consistent picture of this co-evolution. And one of the prominent recurring themes in this geological story (both with and without volcanoes and meteors) is that of species collapse.

Species Collapse in Our Planet's History

It is easy to be flippant about the five massive species collapses that occurred in the distant past. Of course the severe ice age at the end of the Ordovician, 444 million years ago, wiped out 86% of all species! Of course the emergence of land plants during the Late Devonian, 375 million years ago, released nutrients into the ocean, triggering algal blooms that sucked the oxygen out of the ocean, suffocating many sea animals such as trilobites. Of course, at the end of the Permian, 251 million years ago, a perfect storm of natural catastrophes led to a surge in global temperatures, and oceans acidified and stagnated, belching poisonous hydrogen sulfide, leading to the loss of 96% of all species. Of course, at the end of the Triassic, 200 million years ago, for no clear reason 80% of species were lost. And of course, at the end of the Cretaceous, 65.5 million years ago, 76% of all species were lost, perhaps most notably the dinosaurs.

But it makes many people very uncomfortable when they are faced with the undeniable evidence that humans have played a primary role in the transmogrification of the world's ecosystems in far less than 1 million years, and in so doing, are actively contributing to a major wave of species collapse

Species Collapse in Our Planet's History

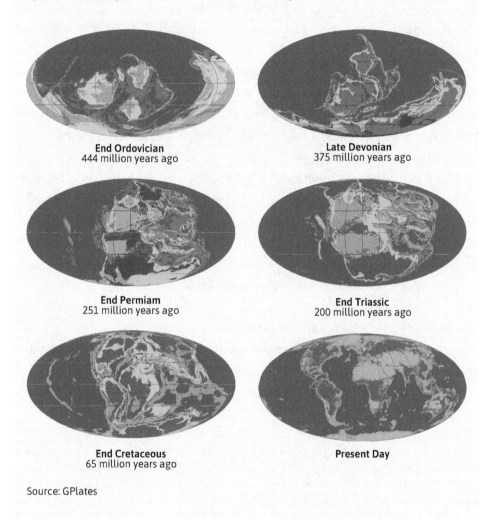

End Ordovician
444 million years ago

Late Devonian
375 million years ago

End Permiam
251 million years ago

End Triassic
200 million years ago

End Cretaceous
65 million years ago

Present Day

Source: GPlates

that is threatening the very carrying capacity of the planet. That's right. Over the past couple of centuries, humanity has steadily eroded the basic ecological underpinnings required to sustain human life on Earth, potentially leading to what some call the Sixth Extinction—also known as the Holocene Extinction or Anthropocene Extinction.[5]

Although the instances of global-scale species collapse in the geological record have a variety of causes, the current wave might best be understood in terms of invasive species. Invasive species are those that are not native to a given ecosystem, whose introduction causes environmental and/or economic harm, as well as potential harm to human health. Though it may have never before happened at a global scale, it is not uncommon for invasive species to destabilize an entire ecosystem, leading to its collapse. A given invasive species can undermine the fundamental factors underpinning an ecosystem through predation or simply by crowding out weaker, yet essential species. It just so happens that in the case of this latest major wave of extinction, we are talking about a single invasive species affecting all of the ecosystems that underpin life on Earth, including human life. In a case of morbid irony, this invasive species is *Homo sapiens*—humanity itself.

Humanity as an Invasive Species

It is time for us to understand humans as an invasive species—*the* invasive species that is undermining the fundamental factors that sustain the global ecosystem. Humanity has a long history of thinking about itself as different

---

[5] An accessible foray into this ongoing cataclysm is Elizabeth Kolbert's book *The Sixth Extinction: An Unnatural History*. However, the term "Holocene Extinction" is used in the vast academic literature that has chronicled the rise of humans as an unprecedented "global superpredator" that has systematically preyed on apex predators while also depleting lower-level species within food webs through hunting and poaching, overharvesting, development, and pollution. The Holocene is the current geologic epoch, which began approximately 11,650 years ago, after the last glacial period, which concluded with the Holocene glacial retreat. This glacial retreat and the associated climate change ushered in major extinctions, which many scientists in part credit to the spread of humans to new ecosystems, as hunter-gatherers. These and other anthropogenic causes of species extinctions have led some to identify this period as the Anthropocene, and the related extinctions as the Anthropocene Extinction. The Holocene-Anthropocene boundary is currently a topic of scientific debate, with some scientists seeing the Industrial Revolution as the beginning of an acceleration in late Holocene extinctions.

from animals. The human consciousness and capacity for moral thinking has spawned all manner of philosophical and religious worldviews that inform us, beginning in childhood, that humans are different from animals. They tell us that humans are not just another species that interacts with ecosystems much like other animals do. Clearly, human capacity for problem-solving, tool building, design, cultivation, manufacturing, trading, writing, philosophizing, governing, and religious belief make us different from animals. But we are loathe to turn that proposition around and think about the cumulative, collective impact that these unique aspects of our species have had on the ecosystems from which we originally evolved and on which our species ultimately depends.

No single data set makes this point definitively. But one data set does provide a stark, eye-opening introduction to the topic that causes you to thirst for additional data that might help you understand the impacts humans have had on specific ecosystems and larger, interdependent Earth systems such as the oceans and the atmosphere. That data set is the spatio-temporal distribution of humanity from 200,000 BCE until now.

Drawn from the archaeological record and rigorous extrapolations based on genetic anthropological methods, this spatio-temporal demographic data set makes several things clear. While the constant flow of newly found humanoid fossils provides endless refinements and adjustments to our understanding of humanity's origins, it is clear that a wide variety of humanoids evolved and spread across the planet since about 2 Ma. But beginning 200,000 years ago, this variety was reduced to four—*Homo erectus* (upright man), *Homo sapiens* (modern humans), the older *Homo neanderthalensis* (Neanderthals), and *Homo sapiens ssp. Denisova* (Denisovans)—if we exclude *Homo florensiensis* ("hobbits"), which enjoyed only a short and geographically isolated moment on Earth from 95,000 to 12,000 years ago. Although *Homo erectus* dated all the way back to 1.8 Ma and roamed Earth until 30,000 years ago, it did so alongside these other hominids, which each came to

Humanity as an invasive species

➜ Human migration, estimated figures in years before present time

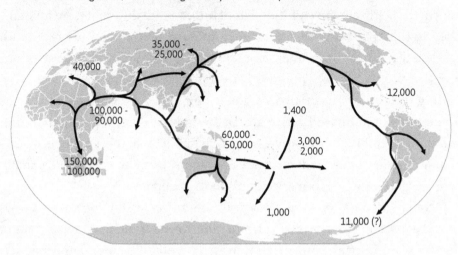

dominate different geographies. Today, humans can rapidly and affordably sequence their DNA, allowing them to discover how much of each hominid their genetic origins comprise.

Homo sapiens, the most recent of these hominids, began its journey in Africa, spreading across Eurasia, Australia, and Oceania, and across the land bridge to the Western Hemisphere. And all along the way, *Homo sapiens* interbred with Neanderthals and Denisovans, and it appears, any other hominid they came across. Best approximations are that this collection of humans, circa 200,000 years ago, stood at about 1 million.

It was 12,000 years ago before some innovative humans, of a global population of roughly 5 million, invented forms of plant and animal domestication, and established geographically fixed settlements, enabling the accumulation of physical, human, and ultimately financial capital with the emergence of expertise, innovation, and increasingly complex divisions of labor. By 7,000 years ago, what we might recognize as some of the first cities were formed, beginning a 9,000-year march of increased urbanization.

The global human population grew at a glacial pace. It took 200 millennia (from 200,000 years ago until 1 CE) to grow from 1 million to fewer than 200 million (estimated at 170 million in 1 CE). After another millennium, in 1000 CE, the human population had risen to only a modest 250 million. And even with the growth of cities and global trade, and the knowledge of Egyptian, Greek, Roman, Indian, Persian, Mesoamerican, and Chinese societies, it took another 800 years before the human population reached 1 billion, in about 1800—partially due to notable dips such as the spread of the Black Death in the 14th century.

It was not until the middle of the 16th century, at a time of Asian dominance and growth, that the population began to steadily rise. Indeed, industrialization in India and China preceded European industrialization, supporting population growth across East, South, and Central Asia, fueled by the trade of the Silk Road. However, the pairing of the European industrial revolution with its scientific revolution created a powerful elixir that reduced mortality and increased longevity steadily across the 18th, 19th, and 20th centuries. Owing to international trade and globalization, these innovations spread far beyond Europe and America to drive rapid population growth all over the globe.

The First Industrial Revolution, all that enabled it and all that came with it, launched humanity onto a path of exponential population growth. The Second Industrial Revolution, met with the dawn of industrialized agriculture and the public health revolution, completed this transformation in population dynamics, leading to nearly vertical population growth beginning about a century ago.

This is not your normal story for an invasive species. *Homo sapiens*, as a species, through collective action, self-consciously shaped its ecological niche over time in ways that enabled near-vertical growth. But it did not do so for the purpose of population growth. Instead, it was in pursuit of "progress." When viewed on a chart or graph or a table of statistics, this growth can easily be viewed as progress. After all, there is an obvious appeal to drops in infant and maternal mortality, and death from gruesome diseases. And humans have a preternatural affinity for innovations that extend the average life expectancy of those who reach adulthood. The 20th century is, in many ways, a story of incredible progress in the quality and longevity of human life. However, when viewed geographically, as humanity spread to previously uninhabited or sparsely inhabited lands, and grew more concentrated over time, one is left to wonder how much of the natural world that has long sustained human life still exists.

This data set, which demonstrates the geographic spread and concentration of humanity over time, leaves little doubt that humanity has somehow transformed into an invasive species over the past two centuries or so, after millennia of treading our planet with a rather light ecological footprint. It prompts us to demand other data sets that could help answer the questions that naturally arise. What was the geographic pattern of deforestation over time? What ecosystems were supplanted by the agricultural domestication and cultivation of various plants and animals, and when? How, where, and when did various species collapse, as humanity spread? What geographies were spoiled by human output and during what time frames? At what point did humanity begin spoiling particular stretches of ocean? And when did society's collective exhaust begin to alter the atmosphere's chemistry in a way that altered climate dynamics, sea level, animal migration, vegetation ecosystems, and ocean chemistry?

Let us pause for just a moment and reflect upon this final question—that of humanity's collective exhaust—and the data sets that might shed light on it. Whereas all the other data sets about our planet are by definition grounded geographically and temporally, the atmosphere and its chemistry are necessarily geographically ephemeral, as the atmosphere's content flows continuously at a molecular level across the globe.

I would argue that the geographically elusive nature of atmospheric data is the single largest factor in our inability to achieve an uncontested consensus on issues of climate change. It is hard enough to communicate humanity's impact on the planet through actions such as deforestation, which is a very tangible, physical process, with concrete visible geographies. When the discussion is reduced to largely invisible gases emanated by various human activities and the largely invisible chemical interactions they have with a largely invisible atmosphere, one can imagine the difficulty in communicating the local, regional, and global impacts of particular gaseous emissions. Many people are inclined to view the number of parts per million of carbon that have been absorbed into the atmosphere and then dissolved into the oceans (another free-flowing global resource) with concern, understanding the serious implications scientists point out. Unfortunately, too many others are quite simply not inclined to "intellectualize" such data. These same people are even less inclined to think through the many ways in which this global chemistry equation is manifesting, or will manifest, in various forms of destruction that impair the carrying capacity of the planet. As such, although we will discuss the choice of geographical data sets that exist, how they unfold over time, and what their implications are for humanity and the planet, we can only afford to put so much effort into teaching people chemistry.

Regardless of this particular complication, there is clearly a common geographic and temporal context in which these larger questions become apparent. When one reflects on those previous wildernesses, which were once largely if not completely unburdened by human settlement, one is certainly compelled to think about what proportion and what specific portions of these ancient wildernesses need to remain intact if our planet and our species are to remain viable over the long term.

Which Half of Earth?

In his recent book *Half-Earth: Our Planet's Fight for Life*, E. O. Wilson asserts that half of Earth's geography must be set apart from humanity's footprint if Earth's imperiled biosphere is to rebound. I am an adherent to the spirit of Wilson's thesis, and I see the book you are reading as a critical though friendly amendment to Wilson's estimate. However, I believe that as one begins to undertake geographically explicit inquiry into the subject, one finds that the challenge is a little more complex than the term "half" might suggest. My thesis that our planet can support only some 3 billion modern humans without incurring long-term ecological debt is driven by an explicitly geographical assessment that suggests that the proportion that must be set aside from humanity could very well be higher, and certainly differs between the terrestrial and the maritime realms.

In Wilson's view, there are coherent contiguous geographies across which complex ecological interdependencies must play out for an ecosystem to function. Earth can continue to sustain humanity only if some collection of these ecological regions are protected or even allowed to recover their historical wildernesses. His assertion that some 50% of these ecosystem geographies must be set aside is somewhat unclear. Does he mean 50% of the planet, in general? 50% of each ecosystem? 100% of half of all ecosystems? Perhaps it is more of an empirical question as to what portion of the world's ecosystems must be set aside in order to produce the biodiversity and ecosystem goods and services needed to sustain humanity. We will explore these questions further in Chapter 6. Regardless, what is clear is that the ecological footprint of human invasion within such coherent geographies militates against the proper functioning of these ecosystems and, at some point, can lead to ecosystem collapse, destroying the ecosystem goods and services on which our human population depends.

It is easy to imagine a distant history when humans lived in harmony with the ecosystems that they inhabited. Indeed, various indigenous populations continue to demonstrate how this can be done, just as their ancestors have done for millennia. Yet ever since humans harnessed the power of fire, some have steadily denuded large swaths of the global landscape, chopping down anything that would fuel the fires they needed for cooking and basic warmth or to provide shelter. Deforestation is not just a modern human process. It is well documented going back thousands of years.[6] And whether forests were felled or other ecosystem forms removed to make way for agriculture and industry, human history, when mapped geographically, showcases a steadily accelerating march toward the annihilation of the ecosystems that have spawned and sustained our very existence.

This is why Wilson advances the thesis that if we do not set aside half of Earth's total geography, safely guarded from the assaults of modern human activity, the very ecological resources required to support our human population could very well collapse. A biologist and not a geographer, Wilson never identifies the specific geographies that he believes should fall within this protected half of the planet. Nonetheless, his work has spawned efforts to digitally map his concepts, such as the Half-Earth Project (www.half-earthproject.org), inspired in part by a related community of scholars who have long argued that "Nature Needs Half" (www.natureneedshalf.org).

If one looks at the latest geographic distribution of *Homo sapiens* as a species, not to mention the vast geographies cultivated to support humanity's insatiable needs, it is clear that humanity would have to decrease its numbers, reduce its per capita ecological footprint, and radically alter its geographic distribution if Wilson's call to action were ever pursued.

---

[6] Extraordinary care has gone into modeling anthropogenic land cover change, going all the way back to the earliest of human settlements. The "KK10 technology scenario," named after work by Jed Kaplan and Kristen Krumhardt in 2010, , is a spatio-temporal data set that has been largely inaccessible outside of scientific circles. This work was preceded by Kaplan et al.'s study on "The Prehistoric and Preindustrial Deforestation of Europe," published in *Quaternary Science Reviews*. To bring the KK10 land use scenario up to present day, scholars have developed a merging strategy to incorporate the postindustrial land use variability of the Historical Database of the Global Environment (known as HYDE).

It is easy for those of us in the developed world to look to lesser-developed regions and demand that their ecosystems be protected at the expense of natural resource extraction that could aid in their development. It is even easier for those countries that currently exercise dominion over these precious ecosystems to point out that the developed world, particularly the United States, displaced and destroyed both ecosystems and indigenous peoples in its quest for prosperity. If America was allowed to benefit from ruthlessly embracing its own Manifest Destiny, why shouldn't developing nations do so?

Indeed, North America was originally home to majestic arboreal ecosystems that were felled in the service of heating and cooking fuels, agricultural land, and ultimately, homes with large, completely unnatural lawns. It was also home to grasslands, deserts, scrublands, wetlands, coastal estuaries, and countless other kinds of ecosystems that stood in the way of "progress" and were engineered out of existence, in part or in whole, to make way for human settlement. The United States, due to its unique role in the history of industrialization, offers an amazing, if disturbing, showcase of the ecological impacts of relentless tooling and machining of the planet's surface in the service of human industriousness.

Interestingly, Americans awoke to this devastation several times and took action. Sometimes, this action was too late to save a particular species. In other cases, action led to preservation and even restoration of critical ecosystems. Concerted action has led to a substantial reforestation of the United States, though too often at the expense of the original type of forest ecosystem. Abrupt changes in pesticide policies have enabled bird and reptile species to rebound from the brink of extinction. Other actions have helped to preserve essential coastal ecosystems, such as mangroves, that serve as essential breeding grounds for all manner of wildlife. Fishery management interventions have helped some fish populations to rebound, or at least provided a temporary stay in their depletion. The United States and many developed nations have taken actions that now incentivize and inspire many in the developing world to halt their own brutal annihilation of the ecosystems that composed their historical landscapes. As such, one would hope that developing countries could see past the developed world's historical ecological transgressions. Unfortunately though understandably, almost everyone in the developing world covets the kinds of prosperity brought by both Western and Eastern economic success and its accompanying ecological devastation.

Perhaps unfortunately for the developing world's economic ambitions, their remaining ecosystems play a much more critical role for our planet, from the perspective of global ecosystem goods and services. The Amazon provides as much as 20% of the oxygen generated for the planet, houses more than 30% of the world's genetic diversity, and produces some 15% of the world's freshwater. Yet Brazil and adjacent nations are fueling agricultural development and human settlement within this vital ecosystem at alarming rates. Central Africa's forests house key iconic species and generate another 10% of the world's oxygen, and thus are commonly referred to as Earth's "second pair of lungs." Sadly, unchecked subsistence farming, informal settlement growth, and burgeoning mineral extraction in the region are increasing the pressure on these ecosystems. The forests of Southeast Asia, particularly in the nations of Indonesia, Malaysia, and the like, are also essential to the planet. Regrettably, global demand for palm oil and tropical hardwoods, not to mention continuous regional population growth, are chewing away at once pristine Asian ecosystems.

To ask these countries to forgo the exploitation of these ecosystems as key sources of economic advance seems unfair, particularly given the United States' historical annihilation of much of its ecosystems—and especially given the economic needs generated by the current Asian, and coming African, population explosion. Yet we must. If the planet is to have enough carrying capacity for even my proposed 3 billion people, these ecosystems must be preserved and even restored. Luckily, when unmolested for just a single generation, these equatorial forests have the ability to rebound and to reclaim that which has been taken from them.

Achieving this, however, demands that modern human forms of ecological devastation be strictly curbed within these geographies. As E. O. Wilson makes pointedly clear, these ecosystems are bundles of interdependencies that have a certain geographical integrity to them. If we expect them to function, they cannot simply be carved apart for particular human uses. Whether transected by transportation corridors, clear-cut for forestry, burned for agricultural land, dammed for energy or recreation, or industrially fished, Wilson makes the case that we cannot reasonably expect the residual portions of these ecosystems to remain resilient in generating the ecosystem goods and services that humanity needs over the long term if it is to survive and thrive. We must begin thinking of how to cordon these ecosystems off from human exploitation so that they can provide us with the sometimes seemingly intangible contributions that they make to the functioning of the planet.

This is not to say that the developed world does not have its own sacrifices to make. The developed world has in many ways retrenched from the geographies that it once worked relentlessly to master, whether as part of Manifest Destiny in the United States or as part of some other equally ambitious cultural project in other countries. There are now farm towns that are depopulating, where their annual industrial-scale agricultural cultivation thwarts historical ecosystems from flexing and reclaiming what was once theirs. Often, this agricultural production occurs only due to subsidies or to serve a larger industrial supply chain that could very well be served in a less ecologically impactful manner. In many places, the historical bucolic ideal of the gentleman farmer or the family-owned farm has become a relic in the dustbin of history, with much family-owned agricultural land now serving as simple extensions of large-scale corporate agribusinesses. In the end, these businesses seek to feed their supply chains and make a profit doing so.

So, when will the market provide opportunities for the ecological reclamation of these historical agricultural lands? After all, the vast majority of these lands were torn from and shorn of their natural ecology only in the past 100 to 150 years. Perhaps, as it relates to the agricultural heartlands of the United States and other developed nations, the 20th century was more of an "ecological anomaly" than a "new normal" that must command the future. What if 10% of agricultural lands were reclaimed to their historical ecologies? What if 50% were reclaimed? With all manner of agricultural innovations, this would not be hard to imagine. Particularly if modern farming, logistical, and retail systems were to reduce agricultural losses from the current 40% to something more like 10%.[7] And, well, if the number of mouths to feed were substantially smaller.

When contemplating which "half" of Earth we might preserve and restore, it is important to consider opportunities to reverse historical ecological devastation that occurred so long ago in such remote places that our current historical consciousness has simply written them off. After all, civilization's emergence in the Fertile Crescent and the Nile River Basin was no coincidence. These places were once lush, if fragile, ecosystems

---

[7] The topic of food waste is often overlooked, particularly when viewed through an economic lens. Economics is focused on how, when, and where food supply and demand for sustenance meet, or where markets fail to meet this demand. Little consideration is typically given to the considerable crop loss during the harvesting, transport, packaging, and retailing processes, unless that loss materially changes prices such that demand cannot be met effectively. Little thought is given to the percentage of Earth's wildernesses that is eliminated and the percentage of Earth's surface that is permanently placed under industrial cultivation, to the exclusion of native species. However, geographically minded scientists who have studied food waste have made it clear that simply reducing waste within our modern food system could enable the reclamation of enormous tracts of wilderness without ever asking modern industrial humans to compromise their lifestyles.

from which relatively small human settlements were carved. These were vibrant floodplains that not only supported these human settlements, but also served as wildernesses occupied by all manner of flora and fauna. They were ecosystems. But they were not the barren desert ecosystems of today. Thousands of years of deforestation for harvesting fuel wood and timber wood, grazing of domesticated animals, and diversion of water in floodplain irrigation projects led to desertification across these regions, which in turn had localized climatological impacts.[8]

In his book *Collapse,* Jared Diamond documented many historical instances of humans who engaged ecosystems and impressed upon them societal demands that simply could not be sustained by the natural growth cycles within the given region. The planet has long been covered by forested desert regions, whether hot or cold in climate, which have very short growing seasons and which take longer than the average human lifetime to regrow after humans take what they want from them. This has meant that humans as a particular form of invasive species, even as a relatively small one, spent much of the first several thousand years of human civilization simply laying waste to flourishing, though delicate ecosystems that offered all manner of goods and services to the planet, in search of fuel, fiber, and building materials.

North Africa is a clear example. Once an abundant forest, which generated a very different climate for the region, and which collected and recycled freshwater across a complex riverine system, we now see it only as the most alienating desert, hostile to most human, plant, and animal

---

[8] Human-induced desertification is a complex process that has recurred repeatedly over the course of human history. Our domestication of plants and animals for agricultural purposes led to widespread land degradation as humans spread across the globe. These processes have only accelerated in modern times. Scholarly treatments of this critical issue abound, as addressed in recent works such as Anton Imeson's book *Desertification, Land Degradation, and Sustainability.*

life.[9] It is entirely unclear that such an ecosystem could be restored, and serious questions could be posed about whether this form of "planetary engineering" is desirable. However, it is clearly a case of human invasiveness undermining the basic underpinnings of a vibrant ecosystem that helped sustain early human civilizations. This, like many of Diamond's case studies, should serve as a cautionary tale about the consequences of humanity's heavy ecological footprint.

But should such historical landscapes be on the list for restoration as part of the quest to set aside half of Earth for nature? A worthy debate.

Before we begin such an accounting, identifying the specific geographies that must be protected in order to ensure ecosystem integrity, we must also engage this thesis that half of Earth must be set aside for nature. Whether Wilson's biodiversity-based Half-Earth formulation or the biogeography-based "Nature Needs Half" formulation, the questions are the same: what to measure and how to measure it.

Is it actually half of Earth's landmass and oceans that we must protect? Is it less than half of the world's landmass that must be protected, but more than half of the world's oceans? Vice versa? Does every equally sized tract of land and sea hold the same ecological significance?

Wilson points out that, at the time of his writing in 2017, the United Nations' World Database on Protected Areas (WDPA) indicated that

---

[9] It is important to note that North Africa's transformation from a lush green ecosystem to the desert that we know today was a complex and discontinuous process. From the archeological record, we know that between 9,000 and 6,000 years ago, a humid phase peaked across North Africa. Some scientists have determined that this African Humid Period was the result of monsoonal climate responses to a periodic slow orbital wobble of Earth that recurs roughly every 20,000 years. Toward the end of the African Humid Period, in about say 3000 BCE, North Africa saw progressive desiccation of the region, leading to widespread depopulation and abandonment of inhabited sites. Although this astronomical phenomenon is well documented by P. B. de Menocal and J. E. Tierney in their 2012 article "Green Sahara: African Humid Periods Paced by Earth's Orbital Changes" in *Nature Education Knowledge,*, even they point out that the nomadic hunter-gatherers populating this region increasingly practiced agricultural cultivation and pastoralism—which required land clearing and deforestation—which no doubt aided the desertification process.

only 15% of terrestrial landmass was under any kind of protection and less than 1.5% of marine environments were protected. On the face of it, that sounds horribly inadequate. So, what percentage of Earth's terrestrial and marine surface should we set aside? What percentage of the planet must be protected in order to provide our planet the basic ecological goods and services needed to sustain itself, and humanity?

Even if Wilson's estimate of 50% is not correct, some version of his thesis is certainly true: Humanity must indeed retrench from its geographically vast enterprise of ecological destruction. Similarly, there is no doubt that the human population must decrease in order to lighten the ecological burden on the planet that sustains our species. But this is the subject of later chapters.

First, it is important to review the history of humanity's impact on our planet through a geographical lens, as humans industrialized the very landscapes from which they evolved.

CHAPTER FOUR

Industrialization of the Global Landscape

Humanity's story on Planet Earth long precedes the story of industrialization, of course. Some 200,000 years of human history saw humanity spread across the planet long before industrialization was even a concept. It is only in the past few centuries that industrialized processes for resource extraction, agriculture, manufacturing, transportation, and urbanization have transformed our global society. Many have focused on the powerful positive benefits that industrialization has had on human fertility and longevity, and the resulting population growth. Little, however, has been said with any kind of geographic specificity about the ways in which industrialization has fundamentally changed the global landscape. Even less has been said about the specific geographies of the ecosystems that sustain life on Earth, and how the process of industrializing Earth's landscape has affected these ecosystems.

It is clear that humanity has launched Earth into its sixth major wave of extinction. Whether one dates its beginning to the retreating of the ice sheets, and the dual impact of climate changes and human predation on the world's megafauna, or to the more recent industrialization can certainly be debated. Regardless, this sixth major wave of extinction clearly kicked into high gear over the 19th and 20th centuries. But over which geographies have these extinctions occurred, and what was the role of industrialization in their extinction?

The earliest days of modern geography, as is the case with many academic disciplines and professional fields of practice, were often in the service of the national ambitions from which the specific geographers hailed. Geography was not just a body of knowledge or a way of thinking. It was a toolbox of concepts and techniques that allowed for the rigorous accounting of any land that became of interest, opening each particular geography to the practical interests of distant nations and the companies that they nurtured. Dr. Isaiah Bowman, an accomplished academic geographer, once the Executive Director of the American Geographical Society, and a founding Director of the Council on Foreign Relations, used the term "conditional conquest" when describing the practical value that his geographical expeditions had for his own nation.[10]

In this sense, the discipline of geography and its professional counterparts in surveying and topographical engineering were complicit in the developed world's various colonial, imperial, and globalization projects. The geography of humanity's footprint, at least as it manifested in the past few centuries, owes much to geographical tools, data, and concepts. That is, the geography of industrialization

---

[10] For many, it is hard to envision a world in which the best maps available lack the detail to navigate a given land, let alone to understand the peoples, ecosystems, and resources it comprises. When modern geographers began mapping the world in the 19th century, long after the Age of Exploration, this was still largely the case outside of Europe, China, and parts of the Muslim world. In his book *American Empire: Roosevelt's Geographer and the Prelude to Globalization*, Neil Smith paints a clear picture of how geographers from several Western nations went around "discovering" distant realms, charting and mapping them to lay them bare for economic, political, and military exploitation. Of course, these lands had long been mastered by their indigenous people, though perhaps not with maps that would have met the cartographic expectations of the likes of Isaiah Bowman. Make no mistake, the industrialization of the global landscape was made possible only through the hard work and daring of geographers trudging across the planet with theodolites and a working knowledge of trigonometry.

owes much to geographers, and in a way, the discipline owes a debt to humanity and our planet for so efficiently laying bare Earth's resources and cultures for exploitation. There is nothing like precise and detailed maps of resources and efficient routes to help move industrialization along.

Although we will certainly continue to discuss the geographical history of humanity's spread and concentration over the past 200,000 years, at this point we will train our attention specifically on the dawn of the Industrial Revolution and the swirl of human activity that it unleashed across our planet over the past three centuries.

What kinds of data might be available to help us understand the impacts of the industrialization of our global landscape? It is important to acknowledge that this chapter could be endless were we to account for every single change to Earth's surface that humans have unleashed at industrial scale. In the interest of brevity, however, we will attempt to provide a rigorous geographical accounting of only five layers of activity.

First, building on the earlier discussion of deforestation in pursuit of fuel, fiber, and building materials, we will focus on industrial-scale extraction of resources from Earth and their use in combustion, in particular. Second, we will reconstruct the major waves of industrialized agriculture that have swept the global landscape and seascape, feeding the growing human civilization at the expense of historical ecosystems. Third, we will explore where and when factories, broadly interpreted, were created to manufacture goods at scale to serve both the needs of the urban populations that grew to support the mass manufacturing enterprise and the needs of others outside of cities. Fourth, we will explore the expansion of infrastructure of various types across the globe, the proximate ecological impacts of different classes of infrastructure, and how infrastructure has enabled access to previously inaccessible lands for settlement and resource exploitation. Fifth, and last, we will view all of this accounting through the lens of urbanization, and particularly through the lens of urban sprawl, suburbanization, and the informal settlements that have come to decimate countrysides around the world—all due to a failure to preemptively plan and strategically invest in sustainable urbanization.[11]

Certainly there are more dimensions to the human ecological footprint that was made so boldly during the past few centuries of industrialization. I am confident, however, that these five dimensions, when addressed geographically, will make the point that humanity has done so much to industrialize the landscape of our planet that we will have to take concerted actions to lighten this footprint if we are to hope to escape the ecological trap that we have set for ourselves.

Fueling Humanity with Extraction and Combustion

Although combustion predates what would generally be considered industrialization, it is important to first explore humanity's original forays into the process. Extraction and combustion have come in many forms over the millennia. Let us start with the basics of humanity's relentless, if inadvertent, effort to denude the global landscape of its forests, beginning

---

[11] To be clear, my formulation is not the only or the best articulation of the human ecological footprint, as I am pursuing a larger thesis. In this book, I distinguish between the industrialization of the global landscape and the emergence of a global geography of persistent waste that this industrialization has precipitated. I do this in order to reframe how we think about and calculate Earth's long-term ecological carrying capacity. But there are many useful treatments of the ecological footprint of humanity that I would encourage everyone to read. One in particular, The Human Footprint: A Global Environmental History, by Anthony N. Penna, offers a global, thematic, and multidisciplinary history of the planet from its earliest origins to its current condition. It is a thoughtful tome that similarly offers a multidisciplinary worldview that draws on the most recent research in geology, climatology, evolutionary biology, archaeology, anthropology, history, demography, and the social and physical sciences. Penna's book offers even more in-depth treatment of the ecological impact of agriculture, urbanization, manufacturing, energy use, and consumption as the global population has grown. Another valuable resource is The Earth as Transformed by Human Action: Global and Regional Changes in the Biosphere over the Past 300 Years, edited by a powerhouse collection of geographers—B. L. Turner II, William C. Clark, Robert W. Kates, John F. Richards, Jessica T. Mathews, and William B. Meyer.

from our earliest days, in search of fuel, shelter, fiber, and arable land that could produce food for cooking with said fuel.[12]

Deforestation in the Quest for Fuel, Fiber, and Shelter

During what is often referred to as "prehistory," humans were hunter-gatherers who hunted and gathered within forests and other vital ecosystems. The use of wood for fuel, fiber, and shelter was modest during this period and did not amount to deforestation of any measurable kind.

Deforestation first appears in the archaeological record thousands of years ago, wherever humanity harnessed organized agricultural methods. Prehistoric Greece and similar historical human settlements hold evidence of deforestation giving way to agriculture, leading to erosion and in many cases ultimately leading to the silting of rivers and harbors. Yet archaeological evidence shows clearly that deforestation accompanied human settlement much earlier, with the rise of organized agriculture and the domestication of grazing animals in North Africa, the Fertile Crescent, China, and Eurasia as far back as 8000 BCE.[13]

---

[12] There is, quite simply, no better book chronicling humanity's complex, interdependent history with our planet's forests than *Deforesting the Earth: From Prehistory to Global Crisis, an Abridgment*, by the late Michael Williams. It is often hard to look at the planet as humanity has transformed it over the past dozen or so millennia, in particular, and understand the wildernesses that our species used to call home. Although dense, his scholarship is a must read for those who care to understand our current struggle with deforestation in the context of our planet's historical forest wildernesses which were, and remain, a critical component of Earth's long-term sustainability.

[13] The field of archeology offers many insights into early human ecological impacts and, in turn, the ways in which ecological destruction often adversely affected human settlements. These studies are necessarily tied to particular geographies and particular moments in time, as in the 2013 article by Tjeerd H. Van Andel et al., "Land Use and Soil Erosion in Prehistoric and Historical Greece" in the *Journal of Field Archaeology*. There are no end of examples of how human-induced ecological destruction in turn negatively affected humanity. As such, we have seen more broadly scoped, quantitative, spatio-temporal analyses going back thousands of years that provide a much-needed synoptic overview of the destruction, such as that addressed in the 2013 article "Used Planet: A Global History" by Erle C. Ellis et al. in the *Proceedings of the National Academy of Sciences*.

Indeed, humanity's impact on the planet's forests was clear and relentless soon after the retreat of the ice sheets saw the return of forests some 16,000 years ago. Since then, humans have been thinning, changing, and eliminating forests in a quest for fuel, building material, fiber, and arable land to shape their economies, societies, and landscapes. Desert regions, both hot and cold, that served as the earliest breadbaskets for humanity, with slow growing seasons, saw quick and permanent losses. Agricultural methods, such as the plow, made it from Egypt all the way to China and India, which saw the first agriculturally induced population explosion and resulting deforestation. Humans migrated northward from Africa and the Fertile Crescent into the newly forested regions of Europe, and over the next 10,000 years would have nearly as much impact on the forests as the glaciers had over 100,000 years.

Many have romanticized the actions that prehistoric peoples took to coexist with nature, even asserting that they were ecologically benign. In truth, they were never in perfect harmony with nature. Rather, early peoples often transformed their landscapes through fire, leading not only to deforestation but also to the extinction of many species across the planet. Other human cultures managed to live within forested ecosystems in a less destructive manner. Often, as an example, people point to the light ecological footprint that Native Americans had on the environment. And compared with the per capita ecological footprint of modern societies, it is undoubtedly true that Native Americans trod more lightly on the land. However, when Europeans first laid eyes on North America, they observed a profoundly disturbed landscape transformed by agriculture and hunting. Although the superstructures of the forests were intact, the forest understories were largely cleared out and curated, for fuel, shelter, and agriculture, fundamentally altering the historical ecosystem. Although this is clearly better than the deforestation driven by other early cultures, as the Native American population soared, the basics of the ecosystems that they lived within were transformed.[14]

Interestingly, just as we think of forests as being inextricably tied to the lives of indigenous peoples, such as Native Americans, forests were similarly tied to the everyday lives of Europeans, as a source of food, firewood, game, and other products, well into medieval times. These forests were wildernesses that held wild animals that struck fear and induced the writing of fairy tales. Dark, dense forests also were thought to be the source of disease.[15] Though there were definitely robust forest ecosystems across Eurasia until the turn of the first millennium, these geographies nonetheless felt the pressure of the plow on their native ecosystems, as it enabled people to cultivate large plots of land. The plow

---

[14] In his groundbreaking book *1491: New Revelations of the Americas Before Columbus*, Charles C. Mann radically changed our understanding of the ecological impact of Native Americans before Western explorers made landfall, beginning with Columbus in 1492. Mann demonstrated how pre-Columbian Indians had deliberately shaped the land around them, with a huge population that did not live as lightly on the land as some would have us believe. Still, by comparison with the dense populations of modern industrialized North Americans, the lifestyles of our indigenous predecessors were significantly lighter on the historical wildernesses that they occupied.

[15] Our world's vast forests were long associated with the miasma theory of disease, which had its roots in ancient times. The word miasma comes from ancient Greek and means "pollution," with the same etymological root as "malaria," which literally means "bad air" in medieval Italian. The theory held that miasma, or poisonous vapors and mists, were filled with particles from decomposing matter (miasmata), which might arise from contaminated water, poor hygienic conditions, or the dank forest floor that was dense with insects and emanated a sometimes smothering foul smell. This miasma theory of disease left humans viewing the world's forests not only as a seemingly inexhaustible source of fuel and fiber, but also as a source of disease that clear-cutting could help ameliorate. This theory was eventually thrown out by scientists and physicians after 1880, replaced by the germ theory of disease proven by Louis Pasteur in 1861, in which specific germs, rather than miasma, cause specific diseases. Interestingly, one of the historical stepping stones that led scientists and doctors to question the miasma theory of disease were garnered from Dr. John Snow's geographic mapping of the Broad Street cholera outbreak in London in 1854. How geographic insights spurred this revolution in scientific understanding is compellingly laid out by Steven Johnson in *The Ghost Map: The Story of London's Most Terrifying Epidemic—and How It Changed Science, Cities, and the Modern World*.

is an ancient technology, going back at least 8,000 years, beginning in 6000 BCE in Egypt, with the Chinese innovation of the iron plow leading to an explosion of land cultivation beginning in 500 BCE across not only China, but also modern-day India. The plow enabled everyday people to become agriculturally productive and led to disproportionate levels of population growth in China and India from these early periods onward, leading to the denuding of forests and other ecosystems in those regions. And in the protection of crops, many species of wild animals were thinned or even extinguished.

Europeans got a later start in agricultural production, population growth, and the resulting annihilation of their forested ecosystems. The Black Death of the 14th and 15th centuries provided some respite from the pressure that medieval Europeans placed on forest ecosystems. But as the population grew, the gradual expansion of agricultural methods and clearing practices gave way to industrial-scale deforestation. The domestication of animals and plant species on such a scale drove the elimination of European forests during the second half of the second millennium, a drive magnified by the harvesting of timber for shipbuilding, urbanization and metal smelting, as industrialization and the move to cities boomed. The iron plow of the Industrial Revolution, credited to the Englishmen Joseph Foljambe in the 1730s and Robert Ransom in 1789, accelerated this process across Europe and its colonies. Then, of course, the American John Deere's steel plow of 1837 unleashed industrialized agriculture on North America and beyond. The final step of industrialization brought mechanization and motive power to the plow and to so many other technologies, so that no ecosystem could possibly stand as an impediment to human "progress." Steam power and then combustion power for felling trees, sawing lumber, and transporting the resulting timber fundamentally changed the pattern and process of deforestation. Clearing of forests became considered an "improvement" to the land. No other ecosystem was immune to this pattern. The period from the mid-16th century to the early 20th century saw similar processes unfold in Australia, New Zealand, Japan, and elsewhere. Wherever Europeans went, the temperate forests of Earth underwent massive, rapid elimination in favor of pastoral landscapes.

The late 19th and early 20th centuries saw a growing awareness of the losses entailed by deforestation. Henry David Thoreau brought attention to the value of forests in the mid-1800s, as a pioneer in the field of nature writing and as a philosopher who saw nature as essential to the human existence. The more philosophical writings of Ralph Waldo Emerson shared this view, as did those of George Perkins Marsh, whose *Man and Nature* made the forceful case that man is on Earth to borrow its natural resources and should responsibly conserve, and if possible, replenish natural resources such as forests. All were widely influential, including famed American conservationists Gifford Pinchot and John Muir.

By the early 20th century, deforestation had raised concerns over a "coming timber famine," which risked the livelihoods and lifestyles of vast portions of American society. Unfettered capitalism and self-interest became viewed as threats of exploitation of the remaining forests in the United States. Pinchot leveraged these concerns to drive the establishment of national forests and the creation of the U.S. Forest Service. Of course, it was not just capitalism that threatened the forests, as the Soviet Union also harnessed the power of modern technology to harvest timber at an alarming rate, to advance its own form of modernism. From the Progressive Era forward, throughout the 20th century, the United States and other nations established national parks, national forests, and all manner of protected areas. Yet the magnitude of the ecosystem annihilation in the second half of the 20th century remained enormous.

Coal Mining as a Temporary Reprieve for Forests

Forests, in many ways, owe their continued existence to the emergence of industrial-scale coal mining. Coal mining provided the fuel that previously was available only by burning trees, wood charcoal, and other biomass. Coal could not replace the wood required to construct shelter, but it did lead to a massive reduction in forest consumption and ultimately created an opportunity for reforestation.

Coal mining and the First Industrial Revolution very much co-evolved, with the abundance and portability of coal as a commodity stoking investment in transportation—first canals and then railroads. Coal has a long history, going as far back as the surface mining of coal in China for household use in 3490 BCE and much later (circa 300 BCE) in ancient Greece for metalworking. But it was the establishment of the first modern coal mine in Scotland in 1575, and then subsequent mining innovations in England, that unleashed a seemingly limitless fuel source for industrialization.

Factories previously powered by waterwheels and thus located on rivers could now be placed across the landscape anywhere that rail could reach, fueled by the coal extracted from Earth. As canals fell out of vogue, railroads propelled humanity across distant geographies, transporting their own fuel, in the form of coal, to the rail network's edges, providing the fuel for their own growth. No frontier was safe from large-scale human settlement, far-flung mining and manufacturing, and the infrastructure to transport timber and agricultural products back to the cities.

By the mid-19th century, no land was beyond the reach of humanity's impending ecological footprint. Coal-fueled infrastructure crossed the biggest rivers, the most perilous mountain ranges, and the most distant wildernesses. Extraction had enabled combustion, which had further enabled endless extraction. The geography of extraction now had a skeleton, in the form of railways and steamboat routes, on which industrialization would soon put more and more meat and fat. And this skeleton soon extended across every continent, save for Antarctica, which indeed it at least reached by ship.

Oil and the Dawn of the Combustion Engine

After coal became dominant, it would be another century before humanity came to recognize the vast abundance of oil trapped beneath Earth's surface, the geographies and geomorphologies that made its occurrence likely, and the technologies and techniques for extracting it.[16] Although surface oil pits have been known for at least 6,000 years, the first oil well was not drilled until 1846, in Baku, Azerbaijan, by Scottish chemist James Young. The oil well widely considered the first in commercial operation was drilled by Edwin Drake in Titusville, Pennsylvania, in 1859. The real turning point during this period was when oil-refining innovations brought us commercial oil refineries that could produce all manner of low-cost oil derivatives, such as kerosene.

For the next century, oil and its various derivative combustibles did not replace coal. They merely helped satisfy the growing human demand for fuel to drive progress. Progress required manufacturing and transport. Progress required heating and lighting.[17] With the dawn of electricity

---

[16] Geomorphology is the study of subtle features in Earth's surface and their relation to larger geological structures. It has long informed exploration efforts for oil and other natural resources. The mapping of geomorphological features stands at the nexus of geography and geology. An outstanding introduction to the emergence of this body of thought is offered in Simon Winchester's *The Map That Changed the World*.

[17] From the 17th century forward, prior to electric street lighting, oil extracted from whales helped fuel streetlights in the developed world. It may be odd to think of whales as a resource that humans extracted from Earth for combustion, but indeed, humans have hunted whales going back to circa 3000 BCE, in part for the extraction of oils for fuel. Of course, few societies had access to whaling as a means of meeting their fuel needs. In the 17th century, whales and their blubber served as a primary source of heating, cooking, and lighting fuels, as well as soap and cooking oil, across Europe, the New World, and Asia. This trade continued into the 20th century, with the dawn of factory ships and the concept of whale harvesting. By the late 1930s, it is estimated that over 50,000 whales were killed each year. The extreme depletion of most whale stocks led the International Whaling Commission to ban commercial whaling in 1986. Still, whale oil never served as a substitute for wood or coal. Instead it offered a liquid combustible, helping to pioneer urban nighttime lighting. Luckily for the whales, ocean biodiversity, and perhaps even humanity's soul, other forms of liquid combustibles—derived from oil—became more dominant by the end of the 19th century.

generation, all of these became beneficiaries of dirty combustion that could be transmitted long distances over high-voltage wires to concentrations of urban dwellers and remote rural populations alike. Moreover, with the revolution in electrical innovation, humans were provided new and varied ways to improve their lot in life, with every new year that passed. Refrigeration. Air conditioning. Elevators. Electric cars. Electric trolleys. Radios. The list goes on seemingly forever.

At about the same time as our landscapes were electrified, oil and its derivatives were paired with the advent of the automobile. While railroads continued to forge new paths across our landscapes, stoked by coal, the automobile encouraged the growth of toll turnpikes and even public roads. Decades after the development of railroads, and even after electric trolley cars, gasoline-fueled buses and cars lowered the up-front capital costs of transporting people and goods to new geographies. Dirt driving paths could be forged with little effort and forged they were, first local and state roads, and then interregional roads. In the United States, the Department of Agriculture funded the expansion and improvement of "farm to market" roads, which connected rural populations to larger cities for transport of their agricultural goods, while also providing rural populations more opportunity. In more long-standing European societies, such roads had existed in some cases back to the Middle Ages or even Roman times. The automobile encouraged their paving. All of this was made possible by the extraction of oil from the planet's crust and by the combustion engine.[18]

---

[18] Although the automobile, with its internal combustion engine, has been the driving force behind the continuous paving of our planet, it was actually the bicycle that helped instigate the modern paved road in America. Carlton Reid, in his book *Roads Were Not Built for Cars: How Cyclists Were the First to Push for Good Roads and Became Pioneers of Motoring*, explores this oft-forgotten wrinkle in the history of our planet's creation of seemingly endless miles of impervious surface. The horse-drawn carriage that preceded the bicycle could manage on well-traveled ruts across an unpaved landscape. As average vehicle speeds and the volume of vehicles increased, the need for smoother, paved roads became imperative for the expansion of the automobile market.

These combustion engines also empowered the Wright brothers, previously bicycle manufacturers, to invent the first powered forms of flight beginning in 1911. The wide-ranging utility of powered flight, and our enthusiasm for it, led to the rapid establishment of a global network of airfields and air communication towers in just the first two decades after its invention. In a way, the ecologically minded could be excited by air travel's ability to move goods and people without scarring Earth with long transportation corridors. However, as time would tell, the aerospace industry would have enormous ecological consequences over the coming decades.

From Forest Holocaust to the Savior of Fossil Fuels

The impact of extraction and combustion on our world's ecosystems cannot be understated. These two processes were literally the fuel that projected modern industrialized humanity across geographies of wilderness that only indigenous populations may have charted before. And despite playing some sort of role in the extinction of earlier megafauna, as noted earlier, those indigenous populations had relatively light ecological footprints compared with modern industrialized humanity.

In this abbreviated overview of the history of extraction and combustion, we see that from its earliest origins, humanity fundamentally changed many of our planet's ecosystems in ways that we have difficulty visualizing in our mind's eye when gazing upon the modern landscapes that we inhabit. Deforestation, at the scale that it happened going as far back as 14,000 BCE, seems to deserve a very different name. If one considers the state of Earth before humans mastered fire, the axe, and the plow, one can only think of what happened afterward as a holocaust. Luckily, during these earliest periods, the human population was small in number, did not span the entire planet, and in some cases, found ways to live in harmony with its natural surroundings. Also, luckily, humans eventually found other things to burn, offering some forests a stay of execution and even allowing some forested ecosystems to regrow. The unfortunate by-products of the industrial-scale burning of these new forms of combustibles were poor air quality, serious impacts on human health, the spoiling of natural ecosystems, and ultimately, sufficiently large-scale emissions to alter our atmosphere and the acidity of our oceans. Yet it was these modern forms of combustibles that enabled so much of what we consider "progress."

The geographical illusion associated with this transition is important to highlight. It is relatively easy to map the ecological impact of humanity burning forest biomass over time, particularly given how absolutely massive it was. It is much trickier to visualize the ecological impact of burning the hidden combustibles extracted from under Earth's surface. The by-products of the continuous, industrial-scale combustion of the past couple centuries have been dispersed into our atmosphere and absorbed by our oceans in ways that are visually indistinguishable. And since we are talking about globally connected bodies of gases and liquids, across which these by-products spread freely, the measurement of the ecological impacts has centered upon scientific observations that are opaque to many laypeople and policymakers alike. However, it took only a couple centuries for the massive accumulation of global-scale combustion wastes in our atmosphere to have measurable geographical impacts. This delayed reaction has spurred enormous debate, as some have leveraged this geographical illusion in order to argue against the existence of ecological impacts of fossil fuels—either out of ignorance or out of malice.

Modernity has required reliable, transportable, energy sources, at a massive scale. But to ignore the ecological impacts of this insatiable thirst for energy is to turn a blind eye to the other half, the darker half, of modernity's equation.

Industrialized Agriculture

Beginning in about 10,000 BCE, humans first domesticated plants and animals. There are competing theories of exactly how, but the when and where are clear.[19] Several different "hearths," distributed widely across the globe, generated these domestications. Rice was domesticated in Southeast Asia in about 10,000 BCE, along the Yangtze River in eastern China. The people of Sub-Saharan Africa domesticated sorghum, yams, millet, and rice in about 5000 BCE.[20] In Latin America, two hearths—in Mexico and in Peru—domesticated beans, cotton, potatoes, tomatoes, and corn between 2000 BCE and 3000 BCE. Southwest Asia was a hearth

---

[19] For a thoughtful treatment of the debates on how agricultural domestication began, it is useful to read James Scott's 2011 Tanner Lecture on Human Values at Harvard University, entitled "Four Domestications: Fire, Plants, Animals, and ... Us." Rather than focusing on the specific instances of plant and animal domestication that occurred independently around the world, Scott contemplates two contending theories of how domestication began. One is the "dump heap" or midden theory, which assumes some kind of sedentary lifestyle in which village or household wastes compost, creating nutrient-rich soil which in turn transforms the various discarded seeds, pits, and cuttings into a garden. The human recognition of this transformation, and organization around it, would have resulted in the birth of agriculture. The second theory has early domestication beginning in rich woodland areas of natural diversity known as Vavilov centers, where inhabitants tended, weeded, and selectively seeded to meet their food consumption needs. Indeed, different crops have different dynamics, and no doubt both dynamics were in play, crop by crop, over different geographies.

[20] The geographer Carl O. Sauer, in his seminal 1968 book *Agricultural Origins and Dispersals—The Domestication of Animals and Foodstuffs*, helped us move beyond the classical view that humanity progressed in stages from hunting to pastoral nomadism to agriculture by introducing the historical hearths in which plants and animals were domesticated, and early theories about how such domestication processes worked. Considerable advancements have been made on this work over the years, through archeological carbon dating and more recently through the genome sequencing of ancient crops. This has made possible detailed mapping of the origins of many crops and their diffusion. Similar techniques have done the same for domesticated animals. All of the dates of domestication are well founded, though approximate due to the methods used and subject to change with the discovery of new samples.

in which many key farm animals, such as cattle, goats, pigs, and sheep, were domesticated in about 8000 BCE. There were multiple instances of dog domestication in Southeast Asia, East Asia, and Europe about 10,000 BCE. The sheep and the goat were domesticated in 7000 BCE in the Fertile Crescent, and pigs, cattle, and chickens in 6000 BCE in Mesopotamia, Mesopotamia and India (contemporaneously), and China, respectively. The horse was first domesticated in Central Asia in about 3500 BCE, and the llama and alpaca around the same time in the South American Andes. Possibly the most recent, the turkey was domesticated in Mesoamerica in about 1 CE.

For thousands of years, these domesticated crops and animals were harnessed at very small scales, spreading slowly from these hearths through trade and nomadic migrations. Nomadic animal herding and subsistence farming were the rule for millennia, and as more coordinated agriculture scaled to the size that could support small urban populations, the ecological footprint of organized agriculture on essential ecosystems was still negligible.[21] Remember, the dawn of agriculture, say 12,000 years ago (that is, in 10,000 BCE), occurred during a period when the human species amounted to fewer than 10 million individuals (perhaps even as late as 5000-3000 BCE). The ice sheets from the most recent Ice Age had just begun receding in 14,000 BCE, after which the northern

---

[21] It is important to put this comment in context. Countless scholarly articles have shown how the spread of early man led to profound ecological change over specific geographies as they brought domesticated plants and animals with them and put these species under active cultivation. One such example is Melinda A. Zeder's 2008 article "Domestication and Early Agriculture in the Mediterranean Basin: Origins, Diffusion, and Impact," published in the *Proceedings of the National Academy of Sciences of the United States of America*. Zeder chronicles how quickly these new domesticated species crowded out endemic species, sometimes to the point of extinction. However, it was a very long time before intensive agricultural cultivation, urbanization, industrialization, and the associated deforestation and ecological destruction reached key ecosystems such as the Amazon and the Congo Basin. Even critical maritime ecosystems were reached only relatively recently.

forests became vital ecosystems.[22] Ten million humans engaged in subsistence farming across six of the seven continents hardly threatened the ecosystems at the core of Earth's carrying capacity.

However, farming soon enabled a substantial increase in population growth, from 10 million in about 3000 BCE to 170 million in about 1 CE. This was an increase from below 0.0005% annually on average over the previous 200,000 years to just over 0.03%. The bulk of that population was in modern-day China and India, in no small part because of the early invention and deployment there of the iron plow and the growing constellation of administratively organized, large-scale agriculture at the core of their societies. While population continued to grow at an annual rate of less than 1/10th of 1%, by 1300 the world's population had reached only 385 million before declining during the spread of the Black Death to about 340 million people in 1400.

The resurgence in population growth after the plague years continued to be strongest in Asia, which experienced an industrial revolution of sorts several centuries before Europe did. The continuous expansion of large-scale, organized agriculture played no small part, with its massive networks of irrigation infrastructure, canals, and trading roads, and the advancement of agricultural technologies. Little wilderness was allowed to persist in the

---

[22] The end of the last glaciation was not a binary moment. For the sake of this narrative, we are focused on the beginning of the northern forests, enabled by the retreat of the glaciers around 14,000 BCE. The Last Glacial Maximum was about 20,000 BCE, as opposed to the Late Glacial Maximum, which was about 11,000 BCE. The Holocene glacial retreat tied to the beginning of the Holocene geological epoch is pegged at 11,700 years ago (9700 BCE). It is commonly thought that due to glacial retreat, humans were able to migrate from the Americas by about 13,000 BCE. New archeological data indicate that this migration may have become possible earlier. It would seem from this record that humans who migrated to the Americas found relatively new postglacial forest ecosystems, while the humans (and other hominids) of Eurasia would have co-evolved with the icy environments and the new forests that succeeded the glacial retreat, as Neanderthals competed with modern humans across Eurasia over the last 60,000 years or so.

face of agricultural development in China, and India suffered a similar degree of ecosystem annihilation due to agricultural expansion.[23]

It would seem that North African and Mesopotamian cultures managed to transform their landscapes from fertile desert ecosystems to largely barren ones significantly earlier. Deforestation, the alteration of river floodplains for agriculture, and the widespread grazing of domesticated animals forever changed these ecosystems.

As discussed in the previous section, clearing trees for agricultural use was a major source of deforestation across Europe. But it was the permanent sedentary lifestyle of cultivation and pastoralization of this land that ensured the forests and other ecosystems would not bounce back to their natural state. And it was the innovations in agricultural technology, techniques, infrastructure, and administrative capacity that intensified this agriculture, at the expense of the historical habitats.

A major part of industrialized agriculture was the establishment of administrative systems that could manage the land and provide for public works such as irrigation and roads. Industrialized agriculture was greater than the solitary farmer and greater than farmers empowered by a plow. It was the complex of skills, methods, procedures, and routines embodied in institutions and administrative capabilities that fostered continuous growth in the agricultural cultivation of land, season after

---

[23] Many focus on the modern ecological destruction that China's and India's massive, growing, and voracious populations are precipitating. However, it is important to recognize that these parts of the world have had relatively large populations for millennia, populations that came very early to industrialized agriculture and other forms of industrialization that long ago annihilated many historical ecological resources. Mark Elvin, in his book *Retreat of the Elephant*, does a particularly good job of presenting how Chinese-style farming, water control systems, and transport infrastructure eliminated the ancient forest wildernesses that were home to elephants and other original species. This history left the China of the 18th century even more environmentally degraded than Europe at that time. Some will note that the enormous European population loss due to the Black Death created space for forest ecosystems to bounce back to some extent.

Spread of agriculture

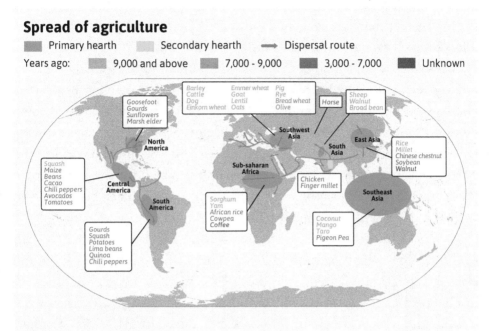

Source: Agricultural Origins and Dispersals: The Domestication of Animals and Foodstuffs

season.[24] These organizational innovations presaged the rise of more productive mechanical technologies—subsequently fueled by petroleum—that later revolutionized industrialized agriculture and managed to destroy far more enormous swaths of the world's natural ecosystems through cultivation.

The case of the American state of Iowa in the 19th century is instructive. In 1833, the first Americans of European descent settled in Iowa. They traveled by rivers, which served as early highways for

---

[24] Thomas Paine, in his essay "Agrarian Justice," asserts that property rights as a social innovation arose from agriculture. Although various property rights regimes have emerged in different ways at different times around the world, agriculture has typically played a prominent role in their formulation. See Chapter 12 for how modern land rights regimes are tied to the industrialization of the global landscape and how their re-imagination would be required if humanity were to rebalance its needs for development with some notion of long-term ecological wellbeing.

explorers, trappers, traders, and ultimately settlers. Pioneers were first allowed to claim new land, under treaty, in the 6 million acre Black Hawk Purchase on the west side of the Mississippi River. John Deere's invention of the steel break-plow in 1837 enabled these settlers to convert the prairies and grasslands into farmland. As over 1 million people settled in Iowa between 1837 and 1870, native wildlife began to vanish due to the drastic loss of the natural landscape to agriculture. The loss of habitat led to the precipitous loss of key animal species, with the last mountain lion in Iowa killed in 1867, bison disappearing in 1870, elk in 1871, black bears in 1876, wolves in 1885, and whooping cranes in 1894.

By 1900, more than 4 million acres of Iowa's original forests had been cut down for development—and this was before the dawn of the gasoline-powered traction engine.[25] By the 1930s, after a couple decades of mechanized farming, Iowa and large swaths of the United States were struck with devastating soil erosion and dust storms. Not only had historical natural habitats that supported animal life been destroyed, but much of the American Midwest struggled to support even human life.

This story is not unique to Iowa, or to America. Humans managed to take industrialized agriculture, whether with fancy machines or with simplistic subsistence farming techniques, one step too far in many places across the globe. Then Iowa, like much of America, and many other places across the globe managed to evolve more sustainable agricultural techniques. Still, the historical habitats remained largely decimated. The industrialization of the global landscape through agriculture has been profound and lasting in its impacts.

---

[25] *John Deere's Steel Plow*, by Edward C. Kendall, recounts the explosive growth in steel plow production (and we must assume, their use), from 10 plows in 1839 to 13,400 in 1857, with annual production growing continually and use accumulating across the increasingly agriculturalized American Midwest.

No story of the rise of modern industrialized agriculture would be complete without a discussion of the emergence of modern techniques for improving soil fertility, and especially the rise of industrial-scale nitrogen fixing after World War II and its application in agricultural fertilizers. Nitrogen is one of the elements that every plant needs in order to grow. Millennia ago, humans awoke to the need to tend to soil fertility. They discovered the power of leaving land fallow for river floodplains to replenish, as well as more formal crop rotation schemes. Humans have fertilized with compost and the application of animal waste, and even human waste. But it was not until the late 19th century that European scientists discovered the specific role that nitrogen plays in plant growth. This happened as part of the revolution in physical chemistry and the rise of industrial-scale chemical engineering in Germany during this same period.[26]

The industrial-scale synthesis of ammonia from nitrogen and hydrogen, two extremely plentiful elements, unleashed an inexpensive, inexhaustible source of agricultural fertility that enabled humanity to feed markedly more people than the natural planet ever could have without this miracle of modern chemistry. It has been said that the expansion of the world's population from 1.6 billion in 1900 to today's 7.5 billion would never have been possible without the industrial synthesis of ammonia. Interestingly, the primary driver for investment in industrial-scale nitrogen fixation in the first half of the 20th century was

---

[26] The dual explosions in human population and industrialized agricultural output have been inextricably linked, and nothing has enabled this pattern more than the revolution in physical chemistry and the dawn of synthetically fixed nitrogen-based fertilizers. The foundations of this transformation have been uncovered by the storied geographer Vaclav Smil in *Enriching the Earth: Fritz Haber, Carl Bosch, and the Transformation of World Food Production*, as well as by the chemist G. J. Leigh in *The World's Greatest Fix: A History of Nitrogen and Agriculture*. Without this chemical innovation taking hold at scale, today's 7.5 billion humans could not be fed. This is perhaps the biggest innovation to demonstrate that Malthus was wrong.

for the production of explosive armaments, with Germany, Japan, and the United States building massive capacity in World War II. At the close of the war, the 10 nitrate plants built in the United States to support the war effort suddenly became an industrial overcapacity that quickly found a home in the agribusiness sector, which expanded rapidly in the postwar United States. With this explosive growth in agriculture in the United States, and around the world as these techniques were exported, came the devastating ecological and health effects of slathering chemicals on our agricultural heartlands, including the poisoning of groundwater, the eutrophication of waterways and water bodies, the creation of ocean dead zones, and even the addition of nitrous oxide (a greenhouse gas 300 times worse than carbon) to the atmosphere.[27] Progressively, throughout the 20th century, nitrogen and the phosphorus that often accompanies it fundamentally changed humanity's ecological footprint for the worse.

The Rise of Manufacturing

Now let us explore where and when factories, broadly interpreted, were created to manufacture goods at scale to serve the needs of the urban populations that grew to support the mass manufacturing enterprise, and the needs of others outside of cities.

The exact definition of a factory is a bit unclear, making its origins a bit vague. If a factory must house machines that aid in the production of goods, then perhaps the equipment in the early blacksmith shops of the Iron Age (about 1000 BCE) would suffice. If not, then the ironworks of the

---

[27] There is bountiful evidence of the historical chemical composition of our planet's atmosphere. The amount of nitrous oxide in our atmosphere has risen 20%, from less than 270 parts per billion to more than 320 parts per billion, since 1750. Fertilizer use and modern agricultural processes are largely responsible for this increase. Nitrous oxide is the third most widespread greenhouse gas, by volume, after carbon dioxide and methane, trapping heat and contributing to global warming while also destroying stratospheric ozone. It has an atmospheric lifetime of 110 years, and the process by which it is removed from the atmosphere destroys ozone.

Han Dynasty (202 BCE-220 CE) no doubt would, as they achieved the scale and sophistication of 19th-century Western ironworks. Certainly the waterwheels that helped grind cereals into flour, perhaps as far back as 500 BCE, would qualify. Indisputably, the Venice Arsenal, founded in 1104, which featured a massive 16,000-person ship production line, utilizing manufactured parts, meets the criteria. The invention of the movable-type printing press in 1439 by Johannes Gutenberg transformed many rooms into factories for manufacturing books.

Nonetheless, it was the First Industrial Revolution, as it began in England, that established what we now think of as the modern factory, which spread across the Old World and the New World like wildfire. John Lombe's water-powered silk mill in Derby was operational in 1721, and many other factories emerged mid-century for the manufacture of pans, pins, wire, and the like. Richard Arkwright is credited with inventing the prototype of the modern factory, with the first successful cotton-spinning factory. This was a new way of organizing labor with machines, in bespoke buildings large enough to house them and optimized for the industrial output of particular goods. Arkwright's factory was widely copied.

Indeed, it was Arkwright's spinning factory design that was the basis of the American industrial revolution, when the young Englishman Samuel Slater convinced Rhode Island industrialist Moses Brown that he could successfully operate the 32-spindle Arkwright apparatus that they could not make work. Slater, who President Andrew Jackson called the Father of the American Industrial Revolution, succeeded in his promise in 1790 and unleashed factory-based manufacturing on the landscape of New World.

It was these factories that created insatiable appetites for cotton and wool to gin or full, spin, and weave. For wood to mill and form into furniture. For metals to smelt, dye, form, and cut. For leather to fashion into shoes, belts, and all manner of finished goods. For wood fiber to transform into paper, and paper to transform into books and other printed materials. And ultimately, for machinery and machine tools that could manufacture new and innovative goods, not to mention more machinery. These machine tools made it possible to supply the railroad industry with the rolling mills, foundries, and locomotive works necessary for its rapid expansion. They made it possible to produce the cast-steel plows and reapers that democratized agriculture and ultimately eliminated the scarcity of food that had long limited population size. They also made possible the successive manufacture of ironclad ships, bicycles, automobiles, and airplanes, which joined the railroads in revolutionizing mobility and making no part of the planet safe from human exploitation.

When widespread electrification became a reality, the location of manufacturing facilities was liberated from streams and rivers. When paired with the railroad and the mobility of coal, manufacturing could be placed at the frontiers to serve frontier populations—or those exploiting frontier populations. Anyone living on the edge of the wilderness could have access to the material fruits of human progress. Factories were now part of the human kitbag for mastering the natural world and "improving" it. Wherever humans might wish to go, there was a clear blueprint for how they could harness the local resources in the service of human consumption and human "progress."

Infrastructure and Its Global Expansion

Humans have been building infrastructure to facilitate their needs for millennia. Infrastructure comes in many forms. Ports and harbors have been built to facilitate seafaring and trade since at least 4,500 years ago, when the Fourth Dynasty of Egypt built one at what is now Wadi al-Jarf, on the Red Sea coast of Egypt. Babylonians dug a 3,000-foot tunnel under the Euphrates River as early as 2160 BCE. Roads were mastered by the Romans and built all across their dominion. Bridges for the transport of goods, aqueducts for the transport of water, sewers for the management of waste, dams for reservoirs, lighthouses for navigation—all of these, and many more forms of infrastructure, have existed since ancient times. It is just that they were the product of know-how developed within only some cultures and were very expensive to construct, and thus occupied only limited geographies. As such, the ecological impacts of infrastructure development were rather limited until the last couple of centuries. Our modern reality has been created not just by the development of more infrastructure, but by the proliferation of new types of infrastructure and the expansion of their geographic reach. Many forms of infrastructure have utterly transformed in their basic character, metamorphosing their interaction with the planet in important ways.

The Roman roads were amazing structures that have demonstrated their permanence over time; however, they never served as impassable corridors that no terrestrial animal could cross. Even the first roads built for automobiles did not have this character. They were long but narrow stretches of pavement with sparse usage that nature largely ignored. Animals, plants, and water treated such linear infrastructure as a minor nuisance. Although these automobile roads certainly helped extend humanity's reach and consequently its devastation to previously remote wildernesses, they hardly stopped nature's natural flows. It was with the dawn of highway corridors in the middle of the 20th century that this all changed across so many of the world's ecosystems. North-south migrations of many terrestrial species all but ceased in some ecoregions, as many old roads gave way to broad and highly trafficked transportation arteries cordoned off by fences, walls, and sound mitigation barriers, and buffered by shoulders denuded of nature. Migration patterns, an essential part of the mating cycle for many animals, were further disrupted as their calls and other biophonic communications were interrupted or drowned out by the continuous drone of transportation-induced noise pollution. The simple paved road evolved into an infrastructural corridor that permanently transected historical ecosystems in ways that somehow escaped the "environmental impact assessments" of the 20th century. The same holds true for other linear infrastructure, such as railways, power lines, pipelines, and canals.

Infrastructure designed for the manipulation and management of natural water flows has had similarly pernicious effects on the functioning of various ecoregions. Twentieth-century hydrological engineering practices proved to have devastating ecological impacts, whether it was the damming and diversion of rivers for hydropower, reservoirs, and water recreation areas, or the creation or maintenance of navigable waterways. Water was seen as a resource for humans, not as a core component in the functioning of the ecosystem within a given region. This is completely understandable, given that the term "ecosystem" was not even coined until 1935 (by Sir Arthur George Tansley, an English botanist).[28] The ecosystem concept took a few decades to become widely accepted, and it took a few more decades before the mainstream civil

---

[28] Tansley coined the term "ecosystem" in his seminal article "The Use and Abuse of Vegetational Concepts and Terms," in the journal *Ecology*. Whereas the words "ecology" and "ecosystem" may be considered close relatives to modern readers, the concept of ecology preceded that of ecosystem by 69 years, when the zoologist Ernst Haeckel defined the new science of ecology (or "Oecologie") in his 1866 book *Generelle Morphologie der Organismen. Allgemeine Grundzüge der organischen Formen-Wissenschaft, mechanische Begründet durch die von Charles Darwin reformirte Descendenz-Theorie* (General Morphology of Organisms: Principles of the Organic Morphological Sciences, Mechanically Grounded in Charles Darwin's Reformulated Theory of Evolution). Haeckel's conceptual innovation (that of "ecology") was published quick on the heels of Darwin's *On the Origin of Species* in 1859. Still, there were conceptual challenges that many scientists struggled with for decades, which Tansley helped resolve with his formulation of the "ecosystem" notion. Key to Tansley's concept was its temporal dimension. He understood ecosystems in evolutionary terms (albeit misapplying the concept of equilibrium to evolutionary systems), as systems that develop over time, becoming more "highly integrated and more delicately adjusted in equilibrium...[as] the normal autogenic succession is a progress towards greater integration and stability." This is to say that we must understand today's ecosystems in historical terms if we are to properly grapple with their long-term viability.

engineering community would embrace the notion that they should "engineer with nature," rather than engineer infrastructure projects that haplessly devastate the ecosystems that the same human beneficiaries need to thrive.

Then there are the infrastructure investments made to support and expand human habitats, from small to large. As we will explore next, in many ways, urbanization is essential to the protection of key ecosystems. However, although urbanization may be a more efficient and sustainable form of human habitat, we should not delude ourselves. Where each city now stands, there was once a pristine ecosystem of some kind. Water flowed through it in distinct ways. Animals enjoyed unique habitats composed of particular plants and soils. These natural habitats generated ecosystem goods and services that offered balance and resilience. Now, those ecosystems are gone, replaced by roads, sidewalks, buildings, power lines, sewers, stormwater management systems, railways, power plants, and the like—along with the occasional patch of "green space." In a way, this is the best-case scenario. Unfortunately, in an increasing number of human habitats, such civil infrastructure has not been developed, leaving humanity's waste by-products to spoil spans of Earth far beyond the specific geography where humans reside. Informal settlements of various sorts tend to bring a level of ecological devastation that is unheard of in modern, well-engineered cities.

It is important to underscore how recent large-scale, ubiquitous infrastructure development has been in the sweep of human history. In the decades preceding the attainment in 1950 of a global population of 3 billion, humanity mastered large-scale engineering methods and by mid-century was unleashing them on the world in pursuit of profit and progress. The expansion of various types of infrastructure across the globe since the early to mid-20th century has had massive ecological impacts as they have provided access to previously inaccessible lands, for both settlement and resource exploitation.[29] In doing so, this infrastructure has transected and dissected ecosystems in ways that have undermined the long-term production of ecosystem goods and services. It has facilitated the spread of invasive species that have wreaked devastation costing countless billions of dollars, and that have compromised the proper functioning of many ecosystems.

Infrastructure, per se, is not the culprit. But it is the prime enabler of human reach across the globe and of the spread of human progress. As such, the geography of humanity's infrastructure investments has, for the past couple centuries, been the pacing variable in Earth's ecological destruction.

Humanity's Creation of an Urbanized World

As humanity has mastered extraction and combustion, agriculture, manufacturing, and infrastructure, it has, in a way, all been in the service of urbanization. Humans brought wood back to their early settlements in

---

[29] Perhaps because of the vast geographical distribution of infrastructure projects around the world, or perhaps due to their relatively recent historical emergence, there is precious little scholarship about the environmental impacts that the last century (or two) of infrastructure development have had on our planet's ecosystems. With no comprehensive treatise on the subject, in what seems to be a true research frontier, we are left to piece together a global narrative from a multitude of case studies. They typically use the language of "environmental degradation" flowing from economic development, where the deployment of large-scale infrastructure is identified as the enabler for the large-scale ecological destruction that accompanies the resulting economic progress. Such studies tend to be focused on specific geographies and conducted by local scholars, or to be broad survey statements by environmental NGOs such as the World Wildlife Fund (WWF). But this is hardly a controversial issue. The World Bank and other leading international financial institutions, led by the Independent Evaluation Group, focus keenly on the "The Nexus Between Infrastructure and Environment," recognizing that if implemented incorrectly, infrastructure investments can lead to economic development that severely undermines ecosystems. What the finance community does not appear to contemplate is whether infrastructure investments might inherently undermine ecosystems, and whether the spread of modern, infrastructure-enabled humanity inherently eliminates ecosystem goods and services.

order to warm, shelter, and cook for their families and neighbors, increasing the probability of survival and enabling the growth of a sedentary population. Humans cleared land for cultivation and grazing in order to provide stable and predictable food sources for their families and neighbors, and to trade with others, also increasing the probability of survival and enabling the growth of a sedentary population. Humans manufactured tools and goods in order to shape the world around them and to augment their capabilities, increasing the probability of survival and enabling the growth of a sedentary population. And humans built infrastructure as the basic physical and organizational structures needed for the operation of their societies, thereby increasing the probability of survival and further enabling the growth of a sedentary population. All of these laid the groundwork for urbanization.

In a way, urbanization is completely unnatural. Humanoids occupied Earth for some 1.2 million years before the first city was founded.[30] Prior to that first city, they wandered a region, or even migrated across the globe, as hunters and gatherers. In truth, that first city was a village by modern standards. Some modicum of fuel, shelter, stable food sources, material things, and infrastructure were required for this first instance of urbanization to occur. But cities as a human innovation caught on quickly. Cities spread across the globe. Cities learned from each other about how urbanization could advance their inhabitants' collective goals. And cities grew in size, complexity, and sophistication. They also grew in terms of their ecological footprints,

---

[30] Whether that first city was Uruk, founded in 4500 BCE in Mesopotamia, or Tell Brak, founded in 6000 BCE in modern-day Syria, or some other place at some earlier date, archeologists have a rather good handle on the earliest human settlements that bear the hallmarks of urbanization. There are many such hallmarks, but the minimum description would be of a large populated urban center of commerce, which has an administration based on a system of laws as well as some kind of regulated means of sanitation.

whether in terms of the wastes that they exuded, the ecosystem area that they displaced, or the species that they affected.

Over the millennia, population growth was not isolated to urban areas. Rural landscapes played host to increasingly dense populations focused on agriculture, whether subsistence farming or more organized and industrialized forms. However, it was urbanization as a social process that enabled humans to shape their immediate surroundings in ways that increased their probability of survival, their hope of longevity, and their quality of life. Urbanization has been key to human progress. Urbanization has also been a driving force behind humanity's rising living standards and unquenchable thirst for Earth's resources. Yet while it is easy to see urban areas as growing scars on Earth, in important ways, urbanization processes offer a pathway toward ecological sustainability for the remaining non-urbanized parts of the planet. If urban sprawl can be contained, sustainable development practices employed, and population curtailed, urbanization will not only dominate the future of humanity, it will also help save our planet from humanity's proclivity for ecological destruction.

Alas, much urbanization is presently not moving toward those goals. The sprawl of informal settlements and suburbanization is driving the growth of urban landscapes and the transmogrification of non-urban landscapes around the world.[31] Although some cities are squarely focused on sustainability, so many simply are not. And urban populations are exploding, both in real terms and as a percentage of population.

---

[31] In the past half-century, urbanization can be understood only while taking into account the plethora of forms of informal settlements—each of which has very different local dynamics fueling its growth. In his book *Planet of Slums*, Mike Davis explores how Brazil's favelas, North Africa's bidonvilles, Argentina's villas miserias, Lima's pueblos jóvenes, and countless specifically named slums such as Beirut's Karantina compose a rich tapestry of informal settlements born of local settlement habits within cities incapable of integrating newly arrived and fast-growing populations into their normal planned patterns of growth. Even under the lowest UN population growth projections, such informal settlements are on track to inflict enormous ecological devastation all over the world by 2050.

Most concerning is the relentless expansion of informal settlements, leading to vast swaths of ecological destruction. Sometime in the late 20th century, humanity fled the countrysides for the cities, leading to all manner of informal settlements without the infrastructure engineering required to make them sustainable. Demands on the periphery landscape intensified, in many places, as more efficient forms of agriculture did not take hold, and cities were fed by subsistence farming or other forms of unsustainable agriculture. Yet the basic services of urban centers further magnified population growth, which spiraled out of control.[32]

To be clear, this is not just the expansion of dense urban cores. This is the continuous and uncontrolled spread of poorly constructed, uninflected, unsustainable structures with little to no infrastructure to enable, sustain, or enhance local living conditions. These informal settlements have decimated countrysides around the world, all due to a failure to preemptively plan and strategically invest in sustainable urbanization. As economists focus narrowly on increasing economic growth and completely fail to see the vast geographies of ecological destruction, one wonders where the "invisible hand" was during this process.

To make it worse, the growth of cities on coasts (as so many are) increased ecological pressures on those precious environments over the course of the 20th century. And to magnify one unintended consequence with another, coastal urbanization increased the exposure of humanity to all manner of natural disasters that converge on littoral zones, such as tsunamis, hurricanes and typhoons, and sea-level rise.

Still, urbanization is a good thing both ecologically and societally, at least as an alternative to other modes of development. Sustainable, smart

---

[32] Countless studies on urbanization trends have been published in recent years. The United Nations' annual *World Urbanization Prospects* report is representative and offers a nice overview of the kinds of dynamics currently in play. Still today, cities around the world are experiencing massive population flows from the countryside and even from distant rural areas.

cities could indeed reduce the per person ecological footprint and lighten humanity's load on the planet. But even sustainable cities are permanent ecological scars that the planet will have to live with, as they have obliterated key ecosystems that will never revive. Urbanization is the pinnacle of the industrialization of the global landscape. The question is whether we can deliberately harness urbanization as a tool in our kitbag as we devise a strategy for saving the planet from humanity, and humanity from our destruction of the planet.

Coda: Glimmers of Enlightened Industrialization

The rise of an ecologically enlightened social movement that has been sloppily called "environmentalism" really took hold in the United States with the publication in 1962 of Rachel Carson's book *Silent Spring*. It sparked a conversation that ended up informing the worldviews of a broad, if loose, confederation of different kinds of groups. This loose confederation has become myriad. It includes urban dwellers who prefer to live in cities but wish for them to have green space and to be sustainable, and who like to vacation in more natural contexts. It includes hunters and game fishermen who are concerned about the sustainability of their prey's ecosystems and about human encroachment on their natural retreats. It includes those who prefer to live more solitary existences in distant rural geographies that are less disturbed by fellow humans. But this confederation has so many more constituent parts. Think of those suburban commuters who live energy-intensive lives that generate enormous volumes of greenhouse gases but are genuinely concerned about climate change as it affects the Arctic. Or those subsistence farmers in the developing world who have few options for a livelihood, yet are aware of the ecological devastation that their lifestyles are wreaking on their native landscapes.

And for the purpose of this discussion, think of those industrialists, and the engineers who work for them, who manufacture the goods that modern society demands, and even yearns for. Many of us yearn for things that have ecological impacts of which we have little to no understanding. These industrialists and engineers may or may not have such an understanding. But if and when they do, they are professionally and sometimes personally invested in a complex business that operates with considerable capital that, once taken, must be paid off. This may be debt that must be serviced over 30 years or capital investment by shareholders that have expectations of returns. It is not so simple to change the terms of this investment or the value proposition of the company, simply because ecological awareness has emerged that did not exist prior to the allocation of capital resources.

That said, it is amazing how much progress many companies have made and what commitments they have made for the future. The nature of particular manufactured goods has been, and continues to be, reconceptualized in order to have a lighter ecological footprint. Manufacturing processes have been reimagined in order to have fewer effluents and fewer by-products. New architecture, engineering, and construction techniques and technologies have been developed that make possible more sustainable human habitats and lighter footprints on ecosystems that have thus far escaped heavy-handed human incursions. Industry has also developed new agricultural techniques and technologies that could lighten humanity's ecological load—though others of these technologies appear to threaten key species and what is left of our historical ecosystems.

In the industrialization of our global landscape, there has certainly been a strange form of progress. Although the accumulated impact on our planet's ecosystems has been epic, new forms of fuel, shelter, agriculture, manufacturing, infrastructure, and urbanization offer the promise of transforming this engineered landscape into something more sustainable. They even offer the promise of reducing the proportion of Earth's ecosystems that must be exploited in order to sustain humanity.

But even with the dawn of enlightened forms of industrialization, the waste—the effluent, exhaust, emissions, sewage, garbage, contaminants, effluvium, and discharge—of humanity seems only to be accelerating, in deeply disturbing ways. This may be the most problematic aspect of humanity's industrialization of the global landscape.

CHAPTER FIVE

The Geography of Humanity's Waste

All species generate waste. Humans, as a species, have always generated waste or by-products of our activity, whether by defecating in a forest or by lighting a fire. Yet as humans have brought forth innovations that let us master our environment and moved beyond hunting and gathering, we have managed to generate new forms of waste that have had all manner of unintended consequences. The geography of humanity's waste as it has evolved is fascinating, mesmerizing, and, quite frankly, horrifying. And like all the other human processes discussed in the previous chapter, humanity's proclivity for waste generation has accelerated aggressively over the past couple centuries and shows no sign of flagging anytime soon.

While Smith, Malthus, and Ricardo wrestled with issues of how the economy could support a population, their debates were silent on the issue of the waste by-products of humans' productive endeavors. Perhaps this was due to the early stages of capitalism that they were witness to. Or perhaps it was because their intellectual pursuits predated concepts of nature and ecology. In this chapter, we will explore some key forms of human waste, their geographical characteristics, their evolution, their ecological consequences, and their implications for the carrying capacity of the planet.

Specifically, we explore humanity's prolific generation of carbon and other greenhouse gases; our creation of ocean dead zones through nutrient surpluses; our release of hard metals, endocrine disruptors, and radioactive materials into the environment and their accumulation in our food chains; our unique predilection for creating all manner of noise pollution that disrupts nature's biophonic processes; and the astonishingly enormous flows of garbage that our global society spews into our oceans, unleashing five continent-size garbage gyres on the wildlife and food chains of our high seas.

Atmospheric Carbon and Our Oceans

Carbon is the form of human waste (perhaps more aptly named exhaust) that everyone talks about these days, as popularized by Bill McKibben and the anti-carbon campaign, 350.org.[33] This advocacy effort helped shine a much-needed light on the impacts of more and more atmospheric carbon on climate change. This effort simplified the pioneering work of the likes of James Hansen, a research scientist at NASA's Goddard Institute for Space Studies, later affiliated with Columbia University's Earth Institute, on how a more diverse set of greenhouse gases have interacted to shape our climate future. There is strong consensus on the impact of humanity's industrially generated carbon on our climate. Only a very narrow segment of the world's

---

[33] 350.org was founded in 2008 by a group of colleagues including journalist and activist Bill McKibben, who wrote one of the first books on global warming for the general public. It was named after 350 parts per million (ppm)—what McKibben, his colleagues, and many scientists assert is the safe concentration of carbon dioxide in the atmosphere. As we have already surpassed this threshold (surpassing 400 ppm in 2018), the global climate change action community has become even more vocal about the need for immediate action, particularly focused on the need for an immediate transition away from the use of fossil fuels. McKibben's 1989 book *The End of Nature* served as a general lament of the threats facing Earth, focused on the greenhouse effect, acid rain, and the depletion of the ozone layer. His 2007 book *Fight Global Warming Now: The Handbook for Taking Action in Your Community* focused attention on the threat of global warming specifically, and served as the battle cry for the 350.org movement, launched that year.

Human-generated carbon

Parts per million, in billions of tons

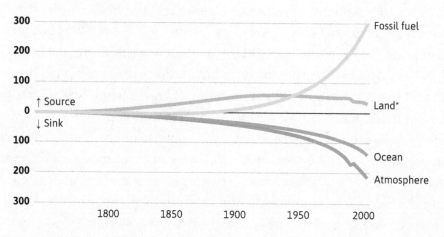

*Land = Fossil fuel + atmosphere + ocean.
Source: World Data Lab

population questions the basics of this atmospheric and ocean chemistry process.

What is perhaps most fascinating is that visible "air pollution" has long faced universal condemnation, particularly when it smells. Visible and olfactive air pollution had its modern-day zenith in the 1970s in the United States and other developed nations, having grown steadily throughout the 20th century with the growth of industrial output. This led to concerted policy action, committed to "clean air." The industrial spoilage of freshwater sources similarly precipitated policy action for "clean water."

At least in America, once most air pollution "disappeared" to the human eye and nose, as a result of the impacts of these policy interventions, popular concern in many ways dissipated. The air seemed fine to most people. So, imagine the reaction when scientists came along to explain why the chemistry of our planet's atmosphere was still highly problematic and indeed could lead us to a cataclysm far worse than the urban smog that had been mitigated in recent years.

This effort started when James Hansen focused his scientific studies, as early as 1980, on "global warming," from an empirical measurement standpoint.[34] His efforts sought to explain variations in net average global atmospheric temperatures, which indeed vary from year to year but still trended heavily toward warming. Through complex calibrations of worldwide temperature observations, Hansen and then others were able to rigorously characterize the net rise in average global temperatures. In his calculations, volcanic eruptions, El Niño occurrences, chlorofluorocarbons, methane, and carbon generated by human processes all interacted within the atmosphere over time to drive these dynamics. And the temperature rise was leading to measurably shorter seasons for generating and preserving Arctic ice, leading to sea-level rise. If the bulk of humanity did not live in coastal zones, perhaps no one would care. But this was alarming. Human exhaust when mixed with other natural processes was magnified in a way that put large portions of civilization at risk.

Hansen's work, famously, led to the creation of the Intergovernmental Panel on Climate Change in 1988. The geographic implications of this work have been shared broadly, with animated maps of our future planet under various scenarios of sea-level rise. It is important to understand that with the current level of atmospheric carbon and ocean heat sinks, and the most optimistic projections for reducing emissions, it is not a question of "if" but "when" the worst-case scenarios will unfold. Rigorous calculations have been made of how ocean water volume will increase due to glacier melt,

---

[34] It is important to note that Hansen was not the first scholar to concern himself with these questions and that, in actuality, these questions have occupied scholars' attention since the Swedish physicist (and later, Nobel Laureate) Svante Arrhenius first inquired about how heat-absorbing gases in the atmosphere affect temperature on the ground, in 1900.

and when juxtaposed with a high-resolution model of Earth's terrain, it becomes clear which portions of our planet will submerge under various ice melt scenarios. Although one can take issue with the exact future dates at which the ocean will engulf various geographies, it is clear that the major coastal cities—where a huge proportion of the world's population and economic value reside—will submerge. Most island nations will disappear. Most of the southeastern United States and northeastern China will exist only as parts of our oceans. Of course, the Arctic will become ice-free sea-lanes. Sea-level rise will destroy countless coastal marine ecosystems as they become too deep to sustain the temperature bands necessary for the historical ecosystems. Ancient animal migration corridors will be permanently eliminated. Of course, this will induce further changes to microclimates around the world, creating even more ecological disruption. The world's physical, biological, and human geography will utterly transform.[35]

Unfortunately, the McKibben 350 ppm rallying cry has served as a distraction from the problem of ocean acidification. It is not just that carbon has been relentlessly pumped into the atmosphere by industrial processes of endless variety and that these will heat the atmosphere, leading to disastrous sea-level rise. It is that the oceans naturally absorb much of this carbon.

As such, when the carbon debate focuses on our atmospheric carbon levels, atmospheric warming, and rising sea levels—all hugely important

---

[35] Estimates from 2017 have the Arctic ice melt streaming 14,000 tons of water per second into our oceans, which will only increase in volume. This rate has tripled since 1986. And although the resulting rising sea levels will have profound impacts on humanity, we need to keep in mind that this epic glacial melt is rapidly moving weight from a concentrated mass to a distributed one that is applying pressure on the tectonic plates, which promises to fundamentally change the pressure equilibrium, leading to increased volcanic and earthquake activity.

and alarming—I would argue that we are distracting everyone from the larger issue of ocean acidification.[36]

Life on Earth depends on the oceans. Ocean acidification has happened before in the geological record, and the result was cataclysmic. The cataclysm that is unfolding now will have a specific geographic extent, albeit under water, and out of sight of most humans. Much has been written about the obvious impacts this is having on the world's reefs. Reef die-off has been accelerating for a while, though it seems to have been largely ignored in popular discourse until recently, as the Great Barrier Reef, our planet's most iconic reef, began its precipitous collapse. However, there is a much larger ocean geography being put at risk by ocean acidification.[37]

---

[36] Despite the atmospheric carbon dioxide that human industrialization has generated, the amount of that gas in the atmosphere is considerably lower than it would otherwise be if not for that which the oceans have absorbed. Unfortunately, the oceans do not simply absorb carbon dioxide in a way that spirits it away. Carbon dioxide reacts with seawater to form carbonic acid. As carbon emissions have increased, more carbon dioxide has entered the world's oceans, creating even more carbonic acid in the water, lowering the water's pH. The ocean pH has been slightly basic for the past 300 million years, averaging about 8.2. Today, it is about 8.1, a drop of 0.1 pH units. This represents a 25% increase in acidity over the past two centuries. As our seawater becomes more acidic, it is able to hold less calcium carbonate, which is a problem of catastrophic proportions since many marine species, including coral, need this calcium carbonate to biologically generate their protective shells and exoskeletons. Without calcium carbonate, this sea life is not able to continue its regular life cycle, and corals erode more quickly than they can generate, depleting the ecosystems that sustain a massive variety of species, leading to their extinction. This might be termed "the other carbon dioxide problem," as Peter Brewer and James Berry called it in their 2008 *Scientific American* article entitled "Rising Acidity in the Ocean: The Other CO2 Problem."

[37] New research by the U.S. National Oceanic and Atmospheric Administration has mapped the areas most vulnerable to ocean acidification, using the distribution of aragonite saturation as a leading indicator. This work shows that ocean acidification is happening on a global scale—showing that large tracts of the Arctic and Antarctic oceans, as well as the upwelling ocean waters off the west coasts of North America, South America, and Africa, are especially vulnerable to ocean acidification. For more, see Li-Qing Jiang et al.'s 2015 study "Climatological Distribution of Aragonite Saturation State in the Global Oceans" in *Global Biogeochemical Cycles* or a popular treatment in *Science Daily* entitled "New Research Maps Areas Most Vulnerable to Ocean Acidification."

Still, the issue of most concern when contemplating the effects of modern industrialized humanity's epic carbon flows remains. That issue should be how the onslaught of atmospheric carbon has been absorbed by the oceans, leading to their acidification, and how this will affect phytoplankton. Phytoplankton, after all, are responsible for the production of the vast majority of the oxygen that humanity and all other animal species depend on. A precipitous die-off of phytoplankton will not only lead to radical change in Earth's oxygen budget. It will also radically change how a fundamental part of the ocean food chain functions. Scientists are just now beginning to get a grasp of how the ocean's changing pH will affect the amount and the mix of various phytoplankton in our oceans, each with different ecosystem functions. This shift in ocean pH will advantage certain groups of phytoplankton over others, and along with ocean and atmospheric temperature changes, will fundamentally change the geography of our oceans' maritime wildlife communities. In short, ocean acidity may very well be the single biggest threat to human civilization.[38]

In light of this perspective, the "climate change" debate is interesting, but perhaps a bit of a distraction. People can deny that climate change is happening or that human industrial and consumer carbon output is the main driver of this change. However, the focus on the scientific esoterica of atmospheric warming and the effects it may have over the coming decades has diverted attention from the obvious peril to our oceans that carbon emissions have induced. Our oceans, after all, cover 71% of Earth's surface, contain 97% of Earth's water, and constitute 99% of the living space on Earth.

---

[38] Hard science has been done on how climate change will alter phytoplankton communities, which will in turn have epic effects on humanity's survival. Studies such as that of S. Dutkiewicz et al. in 2015, published in *Nature Climate Change* as "Impact of Ocean Acidification on the Structure of Future Phytoplankton Communities," make it clear that we face the extinction of many plankton species soon—which should be no surprise after their precipitous decline (some say by 40%) since the 1950s. A lay-accessible article on this topic is available in *MIT News*, entitled "Ocean Acidification May Cause Dramatic Changes to Phytoplankton."

Our failure to focus on the impact of industrial and consumer carbon emissions on our oceans, alone, may very well be humanity's downfall. Perhaps instead we should reframe this debate in terms of the crushing weight of carbon waste that modern industrialized humanity has created in such a short time frame and how it could very well smother much of the life on Earth by first eliminating our ocean's capacity for generating oxygen.

The Metastasizing Geography of Ocean Dead Zones

As if the acidification of the oceans due to atmospheric carbon absorption were not enough to contend with, we also face the very modern emergence of low-oxygen zones in our oceans due to nutrient runoff from our agricultural zones along our rivers and from our coastal cities. There are now over 400 dead zones, the overwhelming number of which are expanding every year.[39] These dead zones sit off the coasts of every continent but Antarctica. Africa's relatively light contribution to coastal dead zones will only increase, as it catches up with global development standards and begins imprinting an equivalent per capita ecological footprint.

Cities generate sewage as well as runoff of effluents from vehicles and industrial activity, and even natural processes can contribute to such hypoxic (lacking in oxygen) conditions. But the clear culprits for the largest dead zones are the nitrogen and phosphorus from agricultural runoff. For instance, the dead zone at the mouth of the Mississippi River, which has recently reached the size of the state of New Jersey, is largely caused by the

---

[39] Up from 49 dead zones in the 1960s, both the number and the size of these zones are growing at an alarming rate. A lay-accessible article on the topic in *Scientific American* entitled "Oceanic Dead Zones Continue to Spread," from a decade ago, paints a rather grim picture—and things have only accelerated. Since then, some of the world's largest dead zones have been discovered in the Bay of Bengal and in the Arabian Sea, previously completely unknown to scientists. Because of the difficulties of remotely sensing such phenomena, we have no idea how many other dead zones are yet to be discovered. All we do know is that the nutrient runoff and other pollutants that cause these dead zones continue to spew into the oceans in increasing volumes.

same agricultural runoff that is causing the algal blooms that are choking life along the Mississippi itself.

Both in the ocean and in freshwater bodies affected by local agricultural and urban development, such hypoxic conditions result from a process called eutrophication, in which an increase in chemical nutrients in the water can lead to excessive blooms of algae that deplete water oxygen levels.[40]

Although low-oxygen zones can occur naturally, their prevalence reflects humanity's influence. Since the 1970s, when dead zones were first detected in the United States' Chesapeake Bay, in Scandinavia's Kattegat Strait, and in the Baltic, Black, and northern Adriatic seas, there has been a continuous climb in the number of hypoxic zones in our oceans, as well as in the total area suffering from hypoxia. These runoff-induced dead zones are only magnified by the rise in atmospheric temperatures, and therefore sea surface temperatures. Human-induced global warming has also had the effect of expanding the otherwise naturally occurring "oxygen minimum zones" across the equator and the Arctic, as warming oceans reduce the amount of oxygen that can dissolve into their waters. Together, these processes have made some 10% of our planet's oceans unlivable for fish and other aquatic organisms.[41] As these dead zones encroach on

---

[40] Again, it is no mystery what causes ocean dead zones. It is so clear that there is a lay-accessible article in *Scientific American* entitled simply "What Causes Ocean 'Dead Zones'?". Perhaps static maps are not enough, and it is simply that we lack a dynamic spatio-temporal data feed that shows these nutrients flowing into water bodies, causing eutrophication through algal blooms and eliminating the oxygen that marine animals need to live.

[41] There is very good reason to believe that this 10% estimate is low. No estimate is available for the annual increase in the overall percentage of ocean consumed by dead zones. Since they are driven by the underlying industrial processes that nourish humanity, there is every reason to believe that the annual rate of increase in total dead zone size will itself increase. This will be a disaster for fish and other marine wildlife, and ultimately for humanity, given the negative feedback loops at the core of this process. The dead zone seafloors turn into biodiversity deserts, leaving little but bacteria that produces methane and hydrogen sulfide, which in turn accelerate this process. A lay-accessible reading on this topic is available at The Conversation website, entitled "Ocean 'Dead Zones' are Spreading – And That Spells Disaster for Fish."

fisheries, and even worse, on mating and spawning grounds, calamity will arise in both the core ocean processes that sustain the planet and the sources of protein for some of humanity's most vulnerable populations.

The intersection of these rapidly changing geographies should be of great concern to everyone. Yet no one has been looking at these various processes geographically, to understand how they might magnify each other. Like a cancer that can metastasize overnight when it reaches or breaches a certain boundary in the body, the geographic spread of dead zones, as they relate to the geographies of other precious processes and resources, should be closely monitored.

The Geography of Persistent Toxicity

Industrialization brought with it the purification of elements and the synthesis of chemical compounds that later proved to be persistent toxic pollutants that accumulate in the environment and in living organisms, including humans. The geography of this persistent toxicity varies with each of their unique histories of innovation and industrial use. Unfortunately, the histories are so complex and intertwined that they have created an intricately woven geographic footprint that has come to cover the globe. As such, the discussion here lacks specific geographical references despite the fact that much is known about the geographies of these modern scourges. This is one instance where maps are clearly more powerful than words.

Nevertheless, we will try to understand the geography of persistent pollutants in terms of three major categories. First, beginning in the 1800s, with the advancements of physical chemistry and the advent of a chemicals industry, a wide variety of chemical compounds arose for all manner of industrial applications, which later in the 20th century were found to disrupt the endocrine systems of living organisms. The use of these endocrine disruptors grew to inundate distinct geographies of our planet. Second, although heavy metals such as mercury, lead, and arsenic have histories of use that go back millennia, only in the past two centuries of industrialization did humans discover and find uses for cadmium, chromium, and thallium. The widespread use of these metallic chemical elements, which have a relatively high density and are toxic or poisonous at low concentrations, led to the accumulation of these persistent pollutants throughout our global food chain. Third, with the dawn of the atomic age, the mining, refinement, and use of radioactive materials, and the disposal of radioactive waste brought us a new class of persistent pollutants that accumulate in the environment, decaying slowly over unfathomably long periods of time. As humanity has partaken of the fruits of industrialization, it has rapidly accumulated these persistent wastes in distinct geographies, undermining the long-term carrying capacity of the planet. Each is worth a bit of your attention.

The Disruption of Core Life Processes

The growth, reproduction, and development of every living organism is in large part governed by hormones. Hormones are controlled by the endocrine system. As such, everyone should be deeply concerned about the industrial production of endocrine disruptors and their geographic spread. Those things that disrupt the growth, reproduction, and development of plants and animals, including humans, represent a form of toxicity that fundamentally undermines the carrying capacity of our planet. Endocrine disruptors emerged with the dawn of modern physical chemistry in the 19th century, and the number and variety of endocrine-disrupting chemicals increased as the physical chemistry revolution unfolded.

These highly toxic chemicals are also known as environmental persistent organic pollutants. They pose significant risks, not only due to their ability to disrupt the proper functioning of life on Earth, but also due to their persistence within the environment. Although each of these chemicals has been manufactured and used for legitimate positive purposes, each has demonstrated persistent, highly destructive effects across the geographies where they have been applied, deliberately dumped, or accidentally released. The "fate and transport" of these persistent contaminants is an active field of study, since each spreads geographically in different ways and at different speeds, depending on how they interact with air, surface water, groundwater, soils, and various living organisms. With the industrialization of the global landscape, the geographies affected by endocrine-disrupting chemicals have become staggeringly large, growing in volume and intensity and continuously spreading as a major threat to environmental and human health. Since hundreds of chemicals are now known to interfere with endocrine systems, it is worth a review of some of the major hormone-altering chemicals, and how their geography threatens our planet and undermines its carrying capacity.

Dichlorodiphenyltrichloroethane, or DDT, was the first compound determined to be an endocrine-disrupting chemical. Though the chemical was originally synthesized in 1874, its insecticidal qualities were discovered more than half a century later, by the Swiss chemist Paul Hermann Mueller in 1939. This colorless, tasteless, and almost odorless chemical was then widely used to control the spread of malaria and typhus. Its positive immediate impacts were so clear that Mueller was awarded the Nobel Prize in Physiology or Medicine. Widely promoted by both government and industry for agricultural, public health, and household use, DDT was used for decades before observations about its environmental impacts accumulated, became well known, and influenced policy. The case of diminishing birth rates and rapid population declines across multiple bird species, perhaps most notably the American bald eagle, was documented in Rachel Carson's instant bestseller *Silent Spring*, which launched the modern environmental movement. DDT was ultimately banned in the United States in 1972 as a result of public pressure.[42]

Many other endocrine disruptors were developed by early physical chemists decades earlier in the 19th century and were put to widespread industrial use. Bisphenol A, dioxins, atrazine, phthalates, perchlorate, perfluorinated chemicals, polybrominated diphenyl ethers, organophosphate pesticides, and glycol ethers are notable examples worth understanding when contemplating how Earth's carrying capacity has been diminished by the industrialization of the global landscape.

Each of these disruptors has its own history and geography as a useful innovation, paired with its unique fate and transport dynamics. Each chemical innovation spread geographically over time and affected the good functioning and health of ecosystems and the plants, animals, and humans within them. This history and geography of innovation and of both postindustrial and postconsumer pollution has stitched a complex quiltwork of persistent pollutants into our global landscape.

Accumulating Heavy Metals

As noted, the human use of heavy metals such as lead, mercury, and arsenic has a long and storied history. Other heavy metals such as

---

[42] Little more than a half-century after *Silent Spring*, the *New York Times* published a piece by the natural scientist Curt Stager entitled "The Silence of the Bugs." In that article, he highlights several studies that pointed to a dramatic decline in flying insect biomass—one study indicating a 75% decline over 27 years. Despite the widespread, though incomplete, moratorium on DDT, humans still demand the broad-based application of pesticides because of insect-borne diseases such as malaria, West Nile virus, and Lyme disease. This is not to mention the desire to eliminate insects as an agricultural nuisance and a nuisance to outdoor living. The industrialization of the global landscape has also reduced the number of insects, as wetlands and the decomposing detritus and duff of forest floors have been replaced by pavement, built structures, and lawns. One unintended consequence has been the elimination of the primary food source for bats, many songbirds, frogs, lizards, and invertebrate arthropods such as spiders and centipedes that prey on insects—leading to precipitous population collapses. Although not all of these pesticides qualify as endocrine disruptors, their continuous application can have similar ecological impacts, with the chemical persistence replaced by the ceaselessness of human procedures designed to rid humanity of the disease and annoyance that comes with the insect densities natural to wilderness.

cadmium, chromium, beryllium, and thallium have been discovered, isolated, produced, and industrially applied more recently. Each has many practical uses, sometimes as pure elements, and sometimes in compounds and alloys. Trace amounts of some heavy metals are required for certain biological processes to function properly. However, these heavy metals also happen to be endocrine disruptors, as well as poisons.

Arsenic famously was the poison of choice for personal and political assassinations from Roman times through the Middle Ages and the Renaissance—long before its industrial utility was discovered. For millennia, lead was used to make coins, cups, utensils, pipes for plumbing, and paint—practical uses that improved human lifestyles while weakening human life. Its smelting contaminated the atmosphere of early Europe for centuries. Mercury, used for thousands of years by the Chinese, Hindus, Egyptians, and Greeks for its beauty, is also a poison. Its symptoms include muscle weakness, loss of coordination, numbness in extremities, rashes, anxiety, memory loss, and loss of the ability to speak, hear, and see. Poisoning from cadmium, one of the more recently discovered heavy metals, causes immediate and irreversible liver, kidney, and respiratory damage. Beryllium poisoning, in contrast, leads to only chronic lung disease among the majority of patients who do not die. Chromium poisoning by inhalation and by oral and dermal contact can lead to immunological, neurological, reproductive, developmental, genotoxic, and carcinogenic effects, as well as death. And in the middle of the 20th century, thallium gained the nicknames "The Poisoner's Poison" and "Inheritance Powder," as a poison of choice over arsenic.

Each of these heavy metals has its own history and geography of use. If we went back a century or so, their geographic footprint would be tied predominantly to the developed world. Now, however, with the globalization of agricultural production, mining, manufacturing, smelting, and fossil-fuel-powered electricity generation, many of the worst cases of toxic pollution are strewn across the developing world.

The database of the United Nations Industrial Development Organization's Toxic Sites Identification Program identifies some 3,000 sites across 47 countries that impose specific public health burdens to over 80 million people. This program is far from finished in identifying such sites, and this number likely represents a small fraction of such sites across the developing world. The U.S. Environmental Protection Agency's National Priorities List, which took 20 years to develop, is a list of 1,300 such sites within the United States alone that require immediate remediation. Given the trends identified by the UN's effort, it appears that the complete survey of such sites would find direct public health impacts on more than 200 million people in the developing world.

It is important to note that this discussion has largely been about the public health impacts of heavy metals. Yet just like endocrine disruptors, these toxic metals have profound impacts on ecosystems. Heavy metals accumulate within the soft tissues of the human body and within the various animals in our food chain as they consume each other. Although each toxic metal has its own environmental impacts, they all lead to plant and animal death, inhibition of growth, and deformity; disruption of photosynthesis; disruption of reproduction; and complex behavioral effects. In addition, or as a result, contamination of environments with such toxins leads to both fewer species and fewer numbers within species.

This is not only true for the original points of pollution. Each toxin has unique fate and transport dynamics, whether by soil, water, air, or organisms, which has led to a geography of ecological spoilage that will persist. And as humanity's industrialization continues to output heavy metals as by-products, the density of their accumulation will only increase across our food chains.

The Half-Life of Radioactive Waste
Radioactive waste is a by-product of nuclear power generation, nuclear weapons manufacturing, and other applications of nuclear fission such as

medical applications. It poses a threat to all forms of life, and its radioactivity decays only slowly over vast periods of time, based on its half-life. Because of this, radioactive waste must be isolated and confined in specially designed disposal facilities for a sufficient period until it no longer poses a threat. In the case of nuclear energy waste, it could take 250,000 years before the materials are safe for contact with organic life-forms. However, due to the use and testing of nuclear weapons, and the occasional nuclear meltdown, considerable amounts of radioactive materials have also been released into the atmosphere.

But for the sake of discussion, perhaps we could suspend disbelief and simply skip over radioactive waste, assuming that it will be properly contained for the next quarter of a million years, or at least put to good re-use. This proposition is patently ridiculous, on its face, but the alternative would require another entire book to contemplate. Nonetheless, we have to keep radioactive waste on our radar when considering human waste.

The Geography of Biophonic Disturbance

Not all wastes take visible, material forms. For example, noise pollution, at first glance, has no tangible qualities.[43] Yet not only can noise be mapped

---

[43] We are explicitly excluding light pollution from this book, despite (or perhaps because of) its pervasiveness. With the dawn of electricity, outdoor lighting quickly became one of the first forms of widespread application. Imagine the view from space in 1802, as William Murdock demonstrated outdoor lighting based on coal gas in Birmingham, England, or in 1878 in Paris when the first electric arc lamp was installed for outdoor lighting. It would have been undetectable on the face of our dark planet. Yet the space-based image of "lights at night" is now a common graphical motif for understanding humanity's inhabitation of Earth. This nighttime light, often called "light pollution," not only prevents humans from seeing the stars while in urban settings, it also disturbs animal interactions with each other and their environment. Artificial nighttime lights on ocean shorelines notoriously affect turtle populations by discouraging females from laying eggs and by disorienting hatchlings. Broadly, many animals are nocturnal, so widespread artificial lighting disrupts their natural biorhythms. Yet the impacts of artificial light are at least largely limited to line of sight, whereas biophonic disturbance can range widely.

geographically, the extent to which it disturbs natural biophony can be measured. Biophony is an obscure concept, but its contamination is disturbing and actively undermines all manner of natural processes that depend on animals signaling through sounds.[44]

Biophony, geophony (non-biological natural sound), and anthrophony (human-induced noise) are the three components of the global soundscape—each with its own distinct geographies. But in many ways, anthrophony (which is sometimes also called technophony, as it is the product of human technology) is a bit of a euphemism. Humanity has a unique predilection toward creating all manner of noise pollution that disrupts nature's biophonic processes. Road noise, airplane noise, ship engine noise, construction noise, factory noise, electrical noise, military test range noise—the list goes on and on. And the geography of this anthrophony has continually spread over the past few centuries as we have industrialized the global landscape. Although there is no definitive global map of anthrophony, the U.S. National Park Service's Natural Sounds and Night Skies Division has generated a nationwide map of sound, specifically in wavelengths audible to humans.

---

[44] The concept of a global soundscape and acoustic ecology is the creation of R. Murray Schafer, a Canadian composer and author of the 1970s book *The Tuning of the World*. His work was followed by that of Bernie Krause, a pioneer of the synthesizer who worked with the Beatles and the Rolling Stones, and contributed to soundtracks of major Hollywood films, who also became the world's most accomplished student and recorder of natural sound. In 2001, Krause wrote an article entitled "Loss of Natural Soundscape: Global Implications of Its Effect on Humans and Other Creatures" for the World Affairs Council. That was followed by his 2002 book *Wild Soundscapes: Discovering the Voice of the Natural World* and 2008 article "Anatomy of the Soundscape" in the *Journal of the Audio Engineering Society*, in which he broke the global soundscape into its constituent parts. This work has been continued by others, such as Almo Farina in the 2013 book *Soundscape Ecology Principles, Patterns, Methods, and Applications*. This work is essential to understanding how humanity has created geographies that are fundamentally hostile to many forms of wildlife. Since then, the practical mapping of the different components of the global landscape has become key to understanding ecosystem wellbeing.

Humans have not only generated sound over a progressively larger portion of the global landscape, over time we have also learned to generate an ever wider variety of sounds—some audible to human ears (20 Hz-20 kHz), some not. Many animals are able to hear frequencies well beyond the human hearing range. Dogs, as everyone knows, hear far beyond the human audible range. Dolphins and bats can hear frequencies up to 100 kHz. Many other animals can hear and are affected by higher-frequency sound. Sound that is lower in frequency than 20 Hz (the normal lower limit of human hearing) is called "infrasound" and is characterized by an ability to cover long distances and get around obstacles with little dissipation. Thus, human-produced infrasound can have impacts on animals that are great distances from the point of origin, such as elephants (down to 14 Hz) and whales (down to 7 Hz). Just because a human hears anthrophonic silence when standing in an ecosystem in no way means that the animals in the same ecosystem are not disturbed and disrupted by human-generated sound.

At any wavelength, anthrophony can intrude upon biophony. When human-generated sound interferes with a segment of the spectrum that is typically used by some set of animals, their communication is disturbed, they cannot make themselves heard, and the information flow that drives natural behavior is compromised. Each ecosystem has distinct biophony; the parts that are audible to human beings have even been mapped geographically. Mating interactions, hunting geolocations, territorial claims—many ecological functions require uninterrupted biophony. DDT killing birds is not the only kind of human waste that could lead to the silent spring that Rachel Carson warned us of.

Soundscape ecology or acoustic ecology—all the biological sounds emanating from a given habitat—will likely continue to go unnoticed, if not unmapped. The ways in which anthrophony—or human-generated sound—affects these ecologies will likely go on being overlooked.

Regardless, this form of human waste only increases as the number of humans grows. And the natural geographies that it affects clearly stretch farther than the human ear can notice.

The Geography of Ocean Garbage

The last form of human waste that we will address in this chapter is the astonishingly enormous flows of garbage that our global society spews into our oceans. Over the course of the past century, humanity has managed to create five floating garbage patches, or gyres, made predominantly of plastic, each the size of a continent. Trapped in swirls generated by our planet's ocean currents, these garbage gyres are growing at an alarming rate and disrupting the wildlife and food chains on our high seas.

No particularly good methodology has been developed for mapping and monitoring the growth of these garbage gyres in terms of surface area (in square kilometers) or volume (that is, area × depth). First discovered in 1997 by Captain Charles Moore, the Great Pacific Garbage Patch was already enormous. In the two decades since, it has gotten bigger (comprising an eastern and a western patch), and further exploration has discovered four other patches in the North Atlantic, the South Atlantic, the Indian Ocean, and the South Pacific.

These garbage patches are composed of a wide assortment of items that would be obviously identifiable to the average person, if only average people wandered the open ocean. Fishing lines and nets, oil drums, plastic bags, plastic and glass bottles, aluminum cans, closed-cell polystyrene foam (Styrofoam™), and all manner of clothing—even diapers—are visible in these garbage patches. The largest proportion of this garbage consists of small pieces of floating plastic that are not immediately evident to the naked eyes of humans, birds, or fish and other ocean creatures. These garbage patches also contain an enormous number of dead sea animals that died after being trapped in the debris, or after choking, starving, or suffering other impairments caused by consuming plastic. Plastic bags can come to look like jellyfish to turtles. Seabirds can mistake floating plastic pellets for fish eggs. Ingesting these plastics can be fatal to sea animals.

Unfortunately, these immense ocean garbage patches only magnify the negative impacts of many of the persistent pollutants discussed earlier, in at least three ways. First, researchers in Japan have shown that when plastics are floating in the oceans, they accumulate and absorb toxic chemicals—particularly hydrophobic chemicals that cling to plastics to escape water—only to be transported into organisms that eat these plastics. These toxic, hydrophobic chemicals include endocrine disruptors such as polychlorinated biphenyls and dichlorodiphenyldichloroethylene (as a common product of the breakdown of DDT), which are derived from pesticides and other human-made substances. Second, as hard plastics decompose in our oceans, they release the endocrine disruptor bisphenol A. These hard plastics are polycarbonates, used to make things like tool handles, CDs and DVDs, eyewear, auto components, medical devices, lighting fixtures, and so on. Although many like to think that these plastics simply litter our environment, unchanged for decades or centuries, in fact polycarbonates biodegrade, releasing bisphenol A into the environment, so that it currently swirls in our oceans, pervades our seashores, and has come to infiltrate our global ocean food chains. Third, the epoxy plastic paint used to seal the hulls of the tens of thousands of merchant ships to protect them from corrosion as they traverse our oceans has dissolved into the oceans, becoming inextricably mixed into these garbage gyres.[45]

---

[45] I am intentionally excluding other pollutants that have been continuously spilled in the oceans for the past century or so, such as oil. They are not technically part of these physical garbage patches, though, as noted, the plastics composing much of the garbage gyres do absorb toxic chemicals as a matter of course. When oil spills, it undergoes weathering over time. A 2017 article by Lee et al. in *Environmental Science: Processes and Impacts* explains that the major toxic contaminants in such spills are polycyclic aromatic hydrocarbons (PAHs) and alkylated PAHs. As they weather, PAHs go through a series of compositional changes which can disrupt endocrine functions. The type of functions affected and associated potencies vary with the type and alkylation status of the PAHs.

Whereas photodegradable plastics can degrade in only six months, other forms of plastic can take hundreds of years to degrade. As such, these garbage gyres will continue to grow. Even if humans manage to vastly reduce the volume of garbage that we spin off of our shores, and even if we vastly reduce the amount of container-ship freight lost to the oceans, these garbage patches will continue to grow over the coming decades. To date, no practical strategies have been devised to diminish their size.

Coda: Prospects for Changing the Wasteful Ecological Footprint of Humanity's Industrialization

As we have industrialized the global landscape, we have not only annihilated historical landscapes and undermined various ecosystems, we have also generated all manner of persistent wastes that have had to accumulate somewhere. The ecological footprint of industrialization thus simply cannot be understood without accounting for the various forms of waste that humanity has spewed across the globe. The rate at which this footprint is growing is awe inspiring. Yet there are signs that this burden could be lightened considerably if collective action were taken to confront the persistent wastes generated by capitalism.

Whether it is carbon and other human exhaust altering the fundamentals of our atmosphere and oceans; nutrients generating ever-growing dead zones in our oceans; persistent wastes such as endocrine disruptors, hard metals, and radioactive materials undermining the health and wellbeing of all living organisms; anthrophonic waste disrupting how animals signal and how ecosystems function; or continent-size garbage patches choking life in our oceans, the geographic footprint of the waste by-products of human progress is growing rapidly.

None of this was ever considered by Smith, Malthus, or Ricardo as they debated the economic dynamics that they believed would guide the industrialization processes that capitalism inspired. The industrialization inspired by Karl Marx's communism and socialism similarly ignored the ecological impacts of the economic systems. There was no room for waste in these equations. And since these great philosophers have dominated so much of our thinking for more than two centuries, our policy debates and policymaking have fundamentally failed to recognize and systematically account for the unintended waste by-products of our economic activity. In so doing, we have been lured into thinking that the explosive human population growth of the past two centuries somehow has not been undermining the carrying capacity of our planet. Well, we were wrong.

Nevertheless, the endless streams of innovations that Smith and Ricardo would expect of modern capitalism do present us with some signs of hope. Recent controversies arising from communities' attempts to rid plastics from their supply chains have created demand for less persistent, even rapidly biodegradable goods. Many of the modern chemicals manufactured for widespread use that have been discovered to disrupt biological processes far from their original intended use are now being phased out of production. The volume and noxious mix of human exhaust has faced more and more scrutiny, with innovations in carbon sequestration, widespread calls for a carbon-neutral society, and the hope that this might also reduce carbon's more toxic exhausts. Innovations have been developed specifically to extract and gather accumulated wastes from the environment, including the massive ocean garbage gyres, so that ecosystems can be restored.

Unfortunately, these glimmers of Ricardian hope must be counterposed with the sheer volume of waste created by the sheer volume of humanity that our planet must somehow support. We face ever-growing populations of ever more affluent people who will only consume more as they move into the global middle class. This means that these innovations alone will not bring our planet and our species back into balance. These technological innovations will only have their desired impact if private sector actors, supported by public sector actors, find ways to accelerate the transformation of commercial supply chains to become fundamentally less wasteful, with less persistent wastes.

In the end, we must find ways to remove these wastes from large swaths of the planet and help define strategies for protecting endangered ecosystems from not only human destruction but also the persistent wastes that we leave behind.

Human Population
in billions

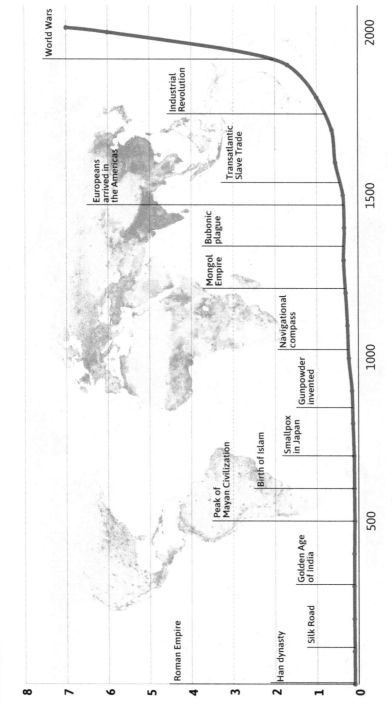

Source: American Museum of Natural History

WWF Global 200
Ecoregions identified as priorities for conservation

■ Marine   ■ Freshwater   ■ Terrestrial

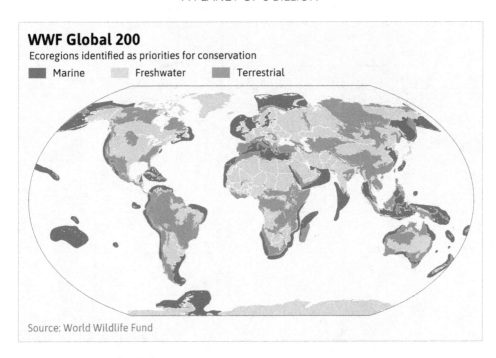

Source: World Wildlife Fund

Nature Needs Half
Progress toward the protection of 50% of the terrestrial biosphere

■ Half-protected   ■ Could reach half   ■ Could recover   ■ Imperiled

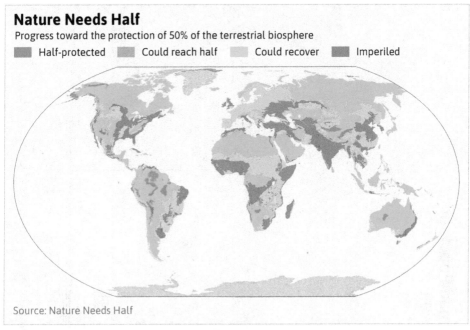

Source: Nature Needs Half

Global Shipping Routes

—— Route

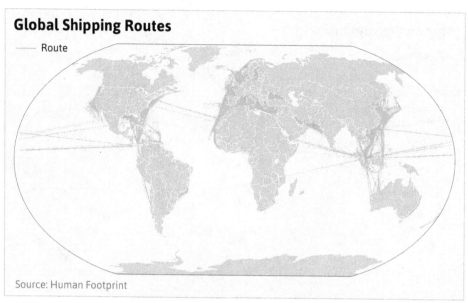

Source: Human Footprint

Last Glacial Maximum Vegetation

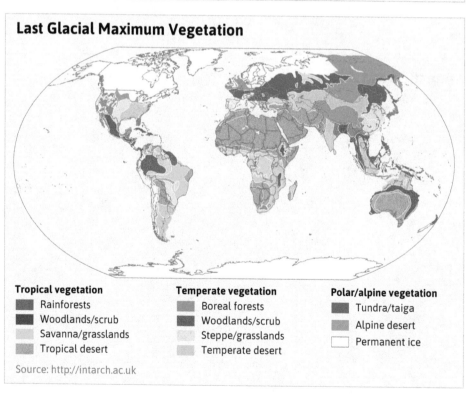

Tropical vegetation
- Rainforests
- Woodlands/scrub
- Savanna/grasslands
- Tropical desert

Temperate vegetation
- Boreal forests
- Woodlands/scrub
- Steppe/grasslands
- Temperate desert

Polar/alpine vegetation
- Tundra/taiga
- Alpine desert
- Permanent ice

Source: http://intarch.ac.uk

Marine Ecoregions

Ecoregion

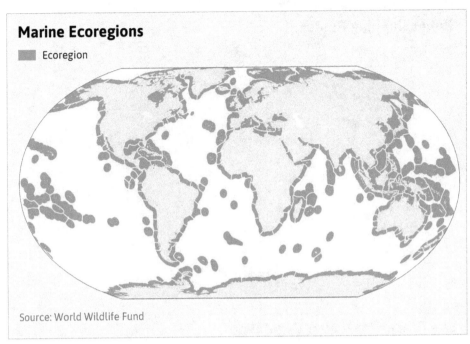

Source: World Wildlife Fund

Aquatic Dead Zones

Dead zones

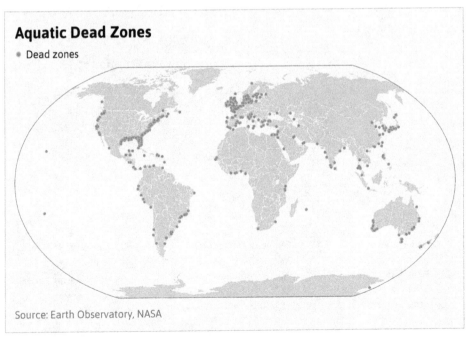

Source: Earth Observatory, NASA

Protected Areas (IUCN categories)

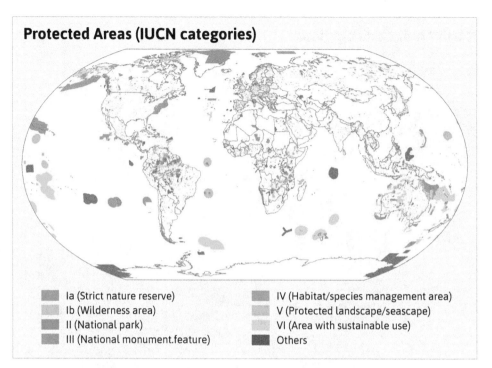

▮ Ia (Strict nature reserve)	▮ IV (Habitat/species management area)
▮ Ib (Wilderness area)	▮ V (Protected landscape/seascape)
▮ II (National park)	▮ VI (Area with sustainable use)
▮ III (National monument.feature)	▮ Others

Ocean Garbage Density

Low ▬▬▬▬▬ High

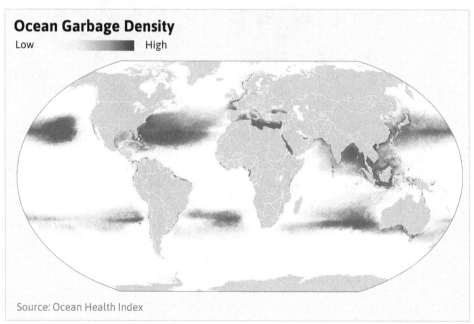

Source: Ocean Health Index

North America: Ecoregions

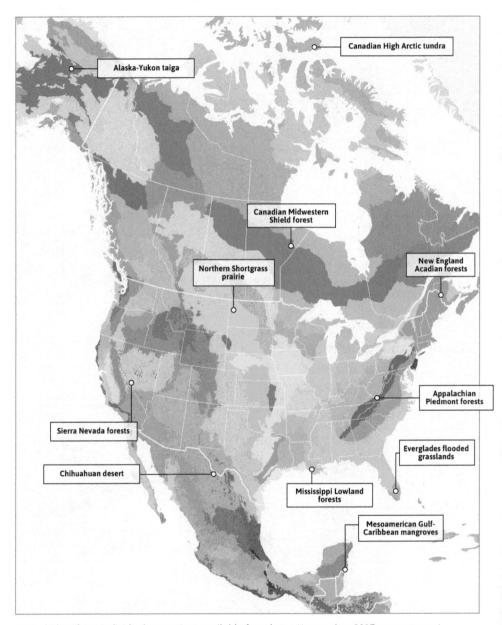

Legend identifying individual ecoregions available from https://ecoregions2017.appspot.com/.

North America: Human Footprints

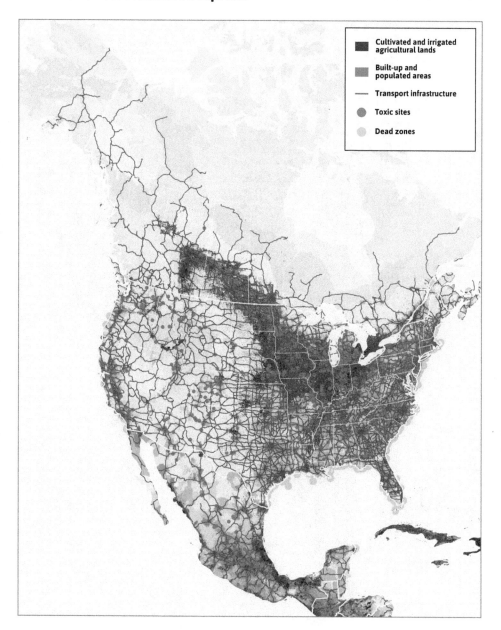

South America: Ecoregions

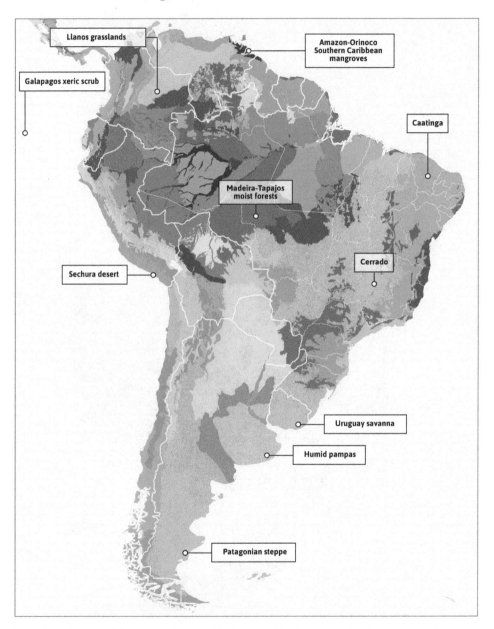

Legend identifying individual ecoregions available from https://ecoregions2017.appspot.com/.

South America: Human Footprints

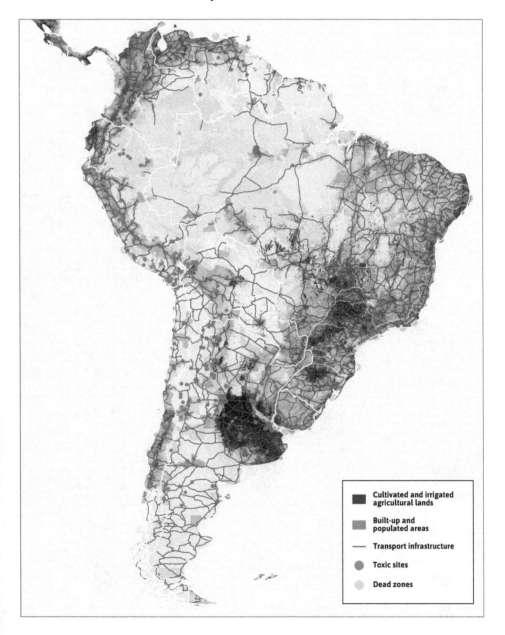

Cultivated and irrigated agricultural lands

Built-up and populated areas

Transport infrastructure

Toxic sites

Dead zones

Europe: Ecoregions

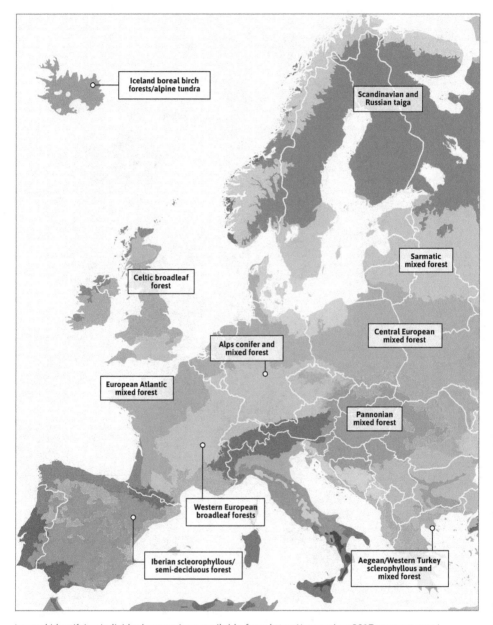

Legend identifying individual ecoregions available from https://ecoregions2017.appspot.com/.

Europe: Human Footprints

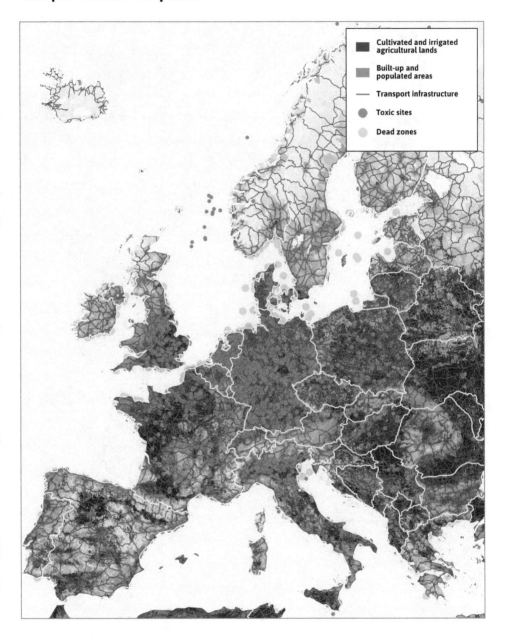

Africa: Ecoregions

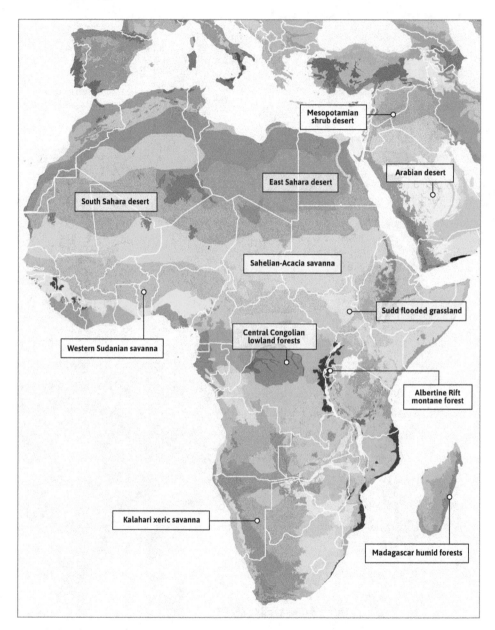

Legend identifying individual ecoregions available from https://ecoregions2017.appspot.com/.

Africa: Human Footprints

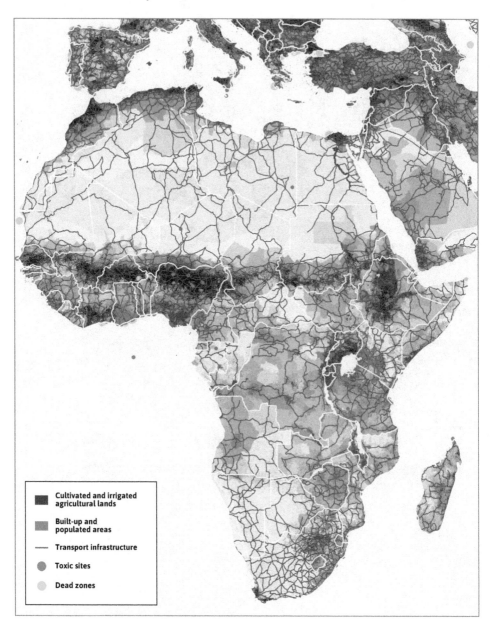

Cultivated and irrigated agricultural lands

Built-up and populated areas

— **Transport infrastructure**

● **Toxic sites**

● **Dead zones**

Asia: Ecoregions

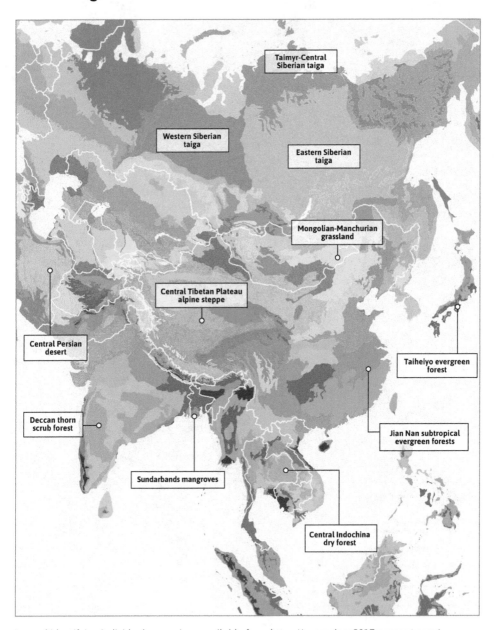

Legend identifying individual ecoregions available from https://ecoregions2017.appspot.com/.

Asia: Human Footprints

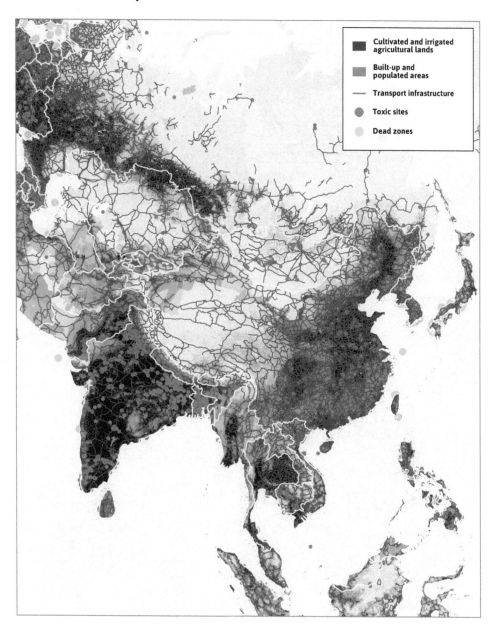

Cultivated and irrigated agricultural lands

Built-up and populated areas

Transport infrastructure

Toxic sites

Dead zones

Oceania: Ecoregions

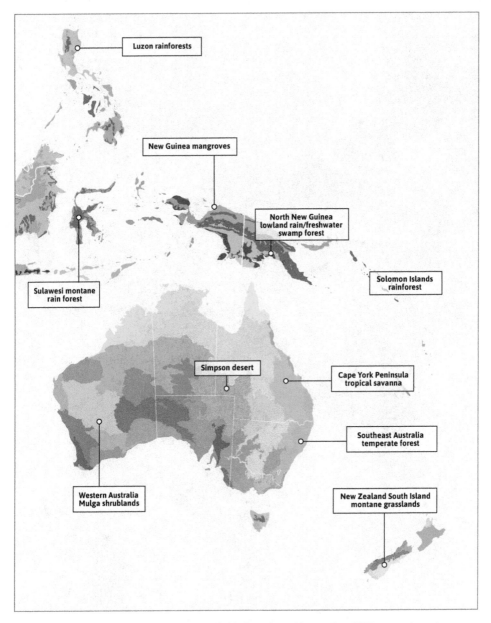

Legend identifying individual ecoregions available from https://ecoregions2017.appspot.com/.

Oceania: Ecoregions

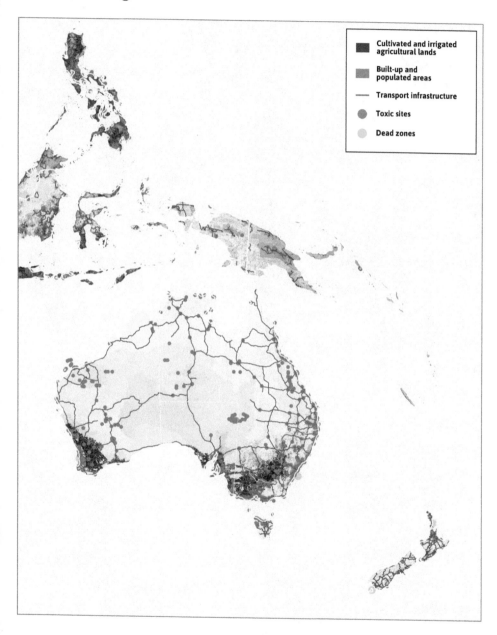

CHAPTER SIX

Defining Protected Areas For Conservation, Preservation And Restoration

---

It is relatively recently in the history of humanity that we have thought to conserve habitats and the natural resources within them. Prior to this awakening, many of the world's dominant cultures, whether Western or Eastern in origin, operated as though pristine ecosystems were simply new lands to be mastered, that natural resources were infinite in their abundance, and that both were meant for humans to harness for their own advancement. The evolution from this mindset to the modern mindset of conservation, preservation, and restoration was long and meandering, and is worth understanding.

The Origins of Conservation Thinking

It is instructive to look first to the American Progressive Era and the role of Teddy Roosevelt and the cohort of public intellectuals and public entrepreneurs who inspired him. Although this may appear to be a very American perspective, the modern global conservation movement is actually deeply rooted in the pioneering efforts of icons of the American Progressive Era such as John Muir and Gifford Pinchot—and Roosevelt himself.

The worldviews of Muir and Pinchot first animated the terms "preservation" and "conservation" that are still used to this day. These very different "land ethics" spawned legal frameworks that helped protect precious natural landscapes and vistas that are now part of the bedrock of American political consciousness. They enabled the creation of over 84 million acres of national parks and national monuments to be protected from all manner of human impacts. However, while it is important to recognize what this accomplishment was, it is also important to recognize what it was not. It was not a system for protecting the ecologically critical parts of the global landscape that sustain human life from encroachment by humanity's industrialization.

John Muir, founder of the Sierra Club, established the preservation movement with the goal of protecting wildernesses from human incursion. He expounded the value of wilderness as part of America's culture and identity. This ethic spawned the National Park Service as a system of wildernesses. Ultimately, this system of national parks became less wildernesses and more preserved landscapes for human recreation. Gifford Pinchot advocated for a conservation ethic, believing mankind inhabits the environment and should serve as stewards of the land's use. Pinchot, the head of Roosevelt's U.S. Forest Service, believed in a scientific understanding of how humans and nature could coexist. It is notable that the U.S. Forest Service sits within the U.S. Department of Agriculture, with a focus on understanding how these forested lands can be harnessed and harvested for human use. These very different land ethics stood in tension with each other, and the relationship of these two contemporaries, Muir and Pinchot, came to a very public climax during the debate between 1908 and 1913 over whether to dam the Hetch-Hetchy Valley in order to supply San Francisco with the water that its growing population demanded.

One additional character is worth knowing in any history of thinking about conservation, preservation, and restoration—Aldo

Leopold. A Pinchot acolyte of sorts, he is the father of so-called "restoration ecology." In actuality, he did not often use the term "restoration." Instead he referred to the practice as "healing the wounded land."[46] Leopold, who attended Yale's School of Forestry, which Pinchot helped endow in 1900, became a thoughtful practitioner of conservation strategies in the 1910s. Prompted by the rise of the automobile in the early 1920s and the increasingly heavy recreational demands it brought to public lands, Leopold was the first to use the term "wilderness" to describe a new kind of preservation that he believed was needed in the national parks and forests. In 1933, he became a professor at the University of Wisconsin and during the Great Depression, he and his fellow professors looked over several hundred acres of eroded farmland on the outskirts of Madison and envisioned its rebirth through restoration.

Although I used the term "land ethic" above to describe the competing worldviews of Muir and Pinchot, the term actually comes from the title of a chapter in Leopold's 1949 bestseller, *A Sand County Almanac: With Essays on Conservation from Round River*. He had concluded that conservation had become a sometimes empty concept that allowed individuals and companies to do what they wanted within the law, while depending on the government to do the rest. Instead, he called for a land ethic that asked humans to enlarge "the boundaries of the community to include soils, waters, plants, and animals, or

---

[46] This is an observation made by Joy Zedler, who served as the University of Wisconsin at Madison's Aldo Leopold Professor of Restoration Ecology and as its Arboretum research director. At the 75th anniversary of the birth of restoration ecology at UW-M, Zedler said, "Ecological restoration really took off here, and Leopold was instrumental in that, but he did not use the term 'restoration' all that often. He wrote about 'healing the wounded land.'"

collectively: the land."[47] He was calling for an ethic in which humans (and specifically, he used the term *Homo sapiens*) would transform their role from that of a conqueror to that of a respectful citizen of this larger ecological community. Such a respectful citizen would, in Leopold's view, engage in active restoration of historical wildernesses, not just in an empty form of conservation.

The modern conservation, preservation, and restoration movements, as embodied in the International Union for the Conservation of Nature (IUCN), are rooted directly in the debates of the American Progressive Era and the actions of Theodore Roosevelt. Roosevelt founded the Boone and Crockett Club in 1887 as a wildlife and conservation organization—named after his heroes Daniel Boone and Davy Crockett, who were prolific hunters. Roosevelt recognized the overharvesting of game across the American wilderness and organized the great naturalists of the day to do something about it. His legacy continued as the Club established the American Committee for International Wildlife Protection in 1930, after spending the 1920s liaising with international naturalists who had organized similar committees.[48] In 1947, the Swiss League for the Protection of Nature called an international conference in which the American Committee participated. This led directly to the establishment of the International Union for Protection of Nature, by UNESCO (the United Nations Educational, Scientific and Cultural Organization), as the first government-organized non-governmental organization (NGO). Renamed the International Union for the Conservation of Nature in

---

[47] Page 239. Leopold's book, often compared with the works of Thoreau and Muir, redefined how we think about the natural world.

[48] Those interested in digging into this rich history can read the 1934 book by John C. Phillips and Harold Coolidge entitled *The First Five Years: The American Committee for International Wildlife Protection.*

1956, this organization has worked to encourage international cooperation in the protection of nature, driving action by national governments and compiling, analyzing, and distributing information about these protected areas. Under this regime, as of 2018, over 161,000 protected areas have been established, representing about 15% of the world's terrestrial surface area, and roughly 15,354 marine protected areas, constituting 7.45% of the world's oceans (calculated by surface area, not volume).[49]

To this day, the very different ethics of protecting wildernesses and conserving managed landscapes, as established by Muir and Pinchot, continue to stand in tension across the globe. Leopold's restoration worldview has proven to be so complex and difficult to implement that there are precious few examples of it. Despite nearly a century of effort, the IUCN process has been able to coax and convince the global community of nations to protect only about 10% of the world's combined terrestrial and ocean surface. Many of Earth's natural habitats are instead under management, so that they can be continuously harvested for human consumption. Still others are exposed to mismanagement or more basic forms of exploitation, with no care for sustainability.

Which lands and seas must be preserved as wildernesses, held apart from human presence, is a continuous debate, with preservation championed by those focused on the vulnerable life stages of various flora and fauna. So too is the debate about conservation. If humans must harness particular geographies for particular productive,

---

[49] Given the rapid growth in protected areas, the latest updates on these statistics are best found at www.protectedplanet.net, which hosts the World Database on Protected Areas and other related data sets.

recreational, or urban habitat purposes, how can we do it in as sustainable a fashion as possible?

What Do We Mean by "Protected"?

The term "protected area" is a formal term defined by the IUCN's World Commission on Protected Areas, which published its most recent guidelines in 2018. The idea inspiring the definition is that we should ensure that entire ecosystems function properly and that their geographical integrity is protected. The IUCN's formal definition states that "A protected area is a clearly defined geographical space, recognized, dedicated, and managed, through legal or other effective means, to achieve the long-term conservation of nature with associated ecosystem services and cultural values."[50]

The IUCN has identified six distinct categories of protected areas, each with different management objectives: (1a) strict nature reserve, (1b) wilderness area, (2) national park, (3) natural monument or feature, (4) habitat and/or species management area, (5) protected landscape and/or seascape, (6) protected area with sustainable use of natural resources. Interestingly, while only about 15% of Earth's terrestrial surface area is formally protected under one of these regimes, and not much more than 5% of Earth's marine surface area is protected, a considerably larger portion of Earth is still largely intact and largely undisturbed. Indeed, there are still places that are mostly wild, with very low human population densities. Many of these are tropical or subtropical moist forests and tundra. That is, they are remote places

---

[50] UNEP-WCMC and IUCN (2016). *Protected Planet Report 2016*. Cambridge, U.K. and Gland, Switzerland: UN Environment Program World Conservation Monitoring Centre and International Union for the Conservation of Nature.

that have not been overwhelmed by human development.[51] Yet what does all this mean? Well, of course, this depends on your object of analysis and how you measure it. I would argue that the concept of ecoregions provides considerable clarity in this effort.

Viewing the World in Terms of Ecoregions

Even among those committed to conservation, preservation, and restoration, there are different disciplinary points of view that inform assessments of our planet's state of wellbeing. Biologists may use a biodiversity lens to develop the boundaries of the biological systems that they wish to protect. Political scientists may use political boundaries. Economists might use administrative boundaries on which economic statistics are kept. Many biogeographers have come to favor the concept of "ecoregions" as a way of dividing up Earth's ecological landscape (and seascape) in a way that respects the territorial integrity of these natural systems. Drawing on biogeographical methods and insights, it is possible to delineate ecosystems composed of particular geographically interdependent distributions of plants and animals

---

[51] A more anthropocentric way of thinking about this transformation is represented by the concept of "anthropogenic biomes," as in the human habitats that have been shaped through millennia of human action, replacing the ancient wildernesses in which humans first evolved. In a wonderful 2008 article in *Frontiers in Ecology and Environment* entitled "Putting People in the Map: Anthropogenic Biomes of the World," Erle C. Ellis and Navin Ramankutty introduce a taxonomy of human-induced biomes: urban, dense settlement, rice village, irrigated village, cropped and pastoral village, pastoral village, rainfed village, rainfed mosaic village, residential irrigated cropland, residential rainfed mosaic, populated irrigated cropland, populated rainfed cropland, remote cropland, residential rangeland, populated rangeland, remote rangeland, populated forest, and remote forest with human populations and agriculture. This left only three categories of wholly natural areas, described broadly as wild forest, sparse trees, and barren land without human populations or agriculture, excluded from the areas that have been anthropogenically remade.

across specific regional extents that have specific climatic, terrain, and hydrological characteristics.[52]

There are related concepts that are important to note if one is to place ecoregions in context. An ecoregion, as an ecologically and geographically defined area, is smaller than a "bioregion" (also called a "biome") and larger than an "ecozone" (also called a "habitat"). A bioregion or biome may span continents, housing many unique ecoregions, and is identified by the type of plant life in relation to temperature and rainfall, though the flora and fauna in different parts of a bioregion or biome often have different genetic lineages. In turn, an individual, distinct ecozone or habitat is likely to reside within an ecoregion, though it may constitute one of many ecological transition zones between ecoregions. The concept of "ecosystem," with which

---

[52] One of the first efforts at delineating ecoregions was a 1976 map of the United States by Robert G. Bailey, a geographer with the U.S. Forest Service (Bailey, R. G. 1976. Ecoregions of the United States (map). USDA Forest Service Intermountain Region. Ogden, Utah. Scale 1:7,500,000). This was followed in 1980 by a Forest Service report entitled "Description of the Ecoregions of the United States." Bailey continued his work with several books: Ecosystem Geography: From Ecoregions to Sites (1996, 2009), Ecoregions: The Ecosystem Geography of the Oceans and Continents (1998, 2014), and Ecoregion-Based Design for Sustainability (2002). Particularly interesting was his use of this biogeographical framework for putting issues of land management, regional planning, and design in context. His discussion of how climate, soils, vegetation, fauna, and human culture interact to define the sustainability of ecosystems is very powerful. His discussions of how ecoregions are changing under human influence are also powerful, and his discussion of how we might use ecological patterns to design monitoring networks and management regimes is a proposal that deserves more attention, and systematic, practical application to ecoregions around the globe. His efforts were complemented by those of James Omernik, of the U.S. Environmental Protection Agency (EPA), whose work (1987, 1995) led to the official ecoregions spatial framework for environmental management embraced in the EPA's official ecoregions data set. Omernik introduced a hierarchical framework for the United States with four levels (I-IV), each progressively more detailed in their delineation of ecoregion borders.

many have at least a passing familiarity, can be either lesser or greater than each of these, geographically. That is, it is a more generic term with less geographic specificity.

Ecoregions cover relatively large areas of land, freshwater, or ocean, each containing characteristic, geographically distinct assemblages of natural communities of flora and fauna. Ecoregions reflect a compromise that delineates a geography that deals with the habitats of as many taxa as possible, but is by no means perfect. Ecoregions, although drawn as distinct geographic regions with a bounding polygon, rarely actually have abrupt edges. Rather, there are ecozones and mosaic habitats at the boundaries between different ecoregions that must be appreciated. These and other habitats within a given ecoregion can actually differ from the larger biome to which an ecoregion is assigned. Ecoregions, as biogeographic provinces, may have their origins in physical barriers (e.g., plate tectonics, geological composition, terrain), climatic barriers (e.g., latitudinal variation, seasonal range, rainfall), or ocean chemistry barriers (e.g., salinity, oxygen levels, temperature bands).

The ecoregions concept has its roots in several decades of thinking. The most modern iteration has its foundations in the work of the World Wildlife Foundation (WWF) and a larger community of scholars, which mapped 867 unique, named, terrestrial ecoregions, as well as 232 marine ecoregions and 426 freshwater ecoregions which, importantly, overlap with the terrestrial ecoregions. The WWF identified the "Global 200," the ecoregions most crucial to the conservation of global biodiversity. The list actually included 142 terrestrial, 53 freshwater, and 43 marine ecoregions, for a total of 238. Many biodiversity-conscious conservationists focus on the tropical rainforests because it is estimated that they house half of Earth's species. An ecoregions methodology, however, offers the foundation for a more comprehensive conservation-preservation-restoration strategy that considers the other half of the world's species also.[53] Ecoregion-based strategies are essential because patterns of biodiversity and ecological processes do not conform to political boundaries.[54] Yet for all the power of the ecoregions concept and the practical process of mapping ecoregions in rigorous, digital forms, there are still some complexities that make it a bit unclear whether it is 50% of the planet that needs protection. Or, at least, it is a bit unclear how one might calculate this 50%.

First, one can analyze ecoregions in different ways. One can take each delineated ecoregion and determine the percentage of its area that is ecologically compromised by human incursions. Human habitats, cultivated agricultural zones, areas of natural resource extraction, and areas curated for recreation can all be subtracted from each ecoregion.

---

[53] Building on the ecoregions work of R. G. Bailey and J. M. Omernik, David M. Olson, Eric Dinerstein, and a community of others affiliated with the WWF added to this body of knowledge from a conservation perspective with a series of articles and books. This began with the 1998 article "The Global 200: A Representation Approach to Conserving the Earth's Most Biologically Valuable Ecoregions." This was followed by two books on the terrestrial (1999) and freshwater (2000) ecoregions of North America, and a 2001 article in *BioScience* entitled "Terrestrial Ecoregions of the World: A New Map of Life on Earth." This global work was aligned with the Level III ecoregion delineations of the U.S. EPA. Dinerstein and his colleagues, through RESOLVE (https://www.resolve.ngo/), a non-profit conservation organization, updated this work in 2017, creating the digital geographic data set at the heart of my arguments in this book. This data set remains aligned with the Level III construct. As such, the Level IV detailing of ecoregions exists only for the United States, in the EPA work done by Omernik.

[54] The global community of biogeographers has applied the ecoregions concept as a tool for studying and conserving biodiversity. There is much work still needed to achieve Level IV granularity in the geographic delineation of ecoregions worldwide. If human history were taught properly, it would prominently feature the ecoregions in which humans evolved, which served as the cradle for human civilizations, and which humanity has progressively annihilated.

However, this simplistic methodology fails to honor the central thesis put forth by E. O. Wilson in his 2016 book *Half-Earth*. Wilson's thesis rests on the notion that ecosystems are complex, highly interdependent systems that must be protected from human incursion. It seems extremely unlikely that an ecosystem that humans have nearly cut in half might sustain anything close to 50% of the ecological interactions or generate 50% of the volume of ecosystem goods and services, and the same mix of goods and services, as when it was wilderness. This logic strains further once one realizes that ecoregions differ in terms of the ecosystem goods and services that they provide to Earth, square kilometer by square kilometer.

Second, mapping the world's ecoregions as they stand today creates a static snapshot, failing to capture the ecosystem boundary shifts that occurred in the past. For instance, the Albertine rift montane forests of Western Uganda have been denuded over thousands of square kilometers over the years, to make way for subsistence farming. The ecoregions methodology, is resilient in how it deals with historical boundaries rather than the apparent boundaries caused by human action. These denuded areas are still recognized as part of the larger historical ecoregion, albeit as parts that have been ecologically diminished. By and large, these ecoregion boundaries have been stable since the end of the last glaciation more than 10,000 years ago, though there is reason to suspect that humans have induced some substantial changes to these ecoregions in some cases, such as the transformation of the trans-Sahara several thousand years ago.

Third, ecosystem interactions are key. Some ecoregions are critical because of the essential ecosystem goods and services that they provide to other ecoregions. These goods and services could flow from water, plants, animals, or even the weather and microclimates that an ecosystem may induce for its neighbors. Water from one ecoregion may be essential to the full functioning of another ecoregion.[55] Flora within one ecoregion may generate nutrients or microclimates that adjacent ecoregions depend upon for day to day, or seasonal, functioning.

Fauna present an additional set of interactions. Not only may animals from an adjacent ecoregion play an essential role in the proper function of a neighboring ecoregion, but migratory animals from distant ecoregions, even on another continent, may be essential to both ecoregions, as well as ecoregions along their migratory path. This certainly holds true for migratory birds, migratory land animals of many kinds, and migratory sea animals such as whales, sea turtles, and some sharks. The migrations of ocean wildlife require contiguous maritime wildernesses that offer appropriate nutrients on the long journeys while protecting the migrants from human molestation. The migrations of land animals require coherent migratory corridors which can sustain them and over which they can move without impediments. Migratory birds require airspace corridors in which they can travel without destruction by human technologies (whether a shotgun, a wind turbine, or DDT), and where their age-old migratory waypoints are protected.

As such, when one takes seriously E. O. Wilson's focus on interdependence, the term "protected" takes on new meaning. Whereas some may recoil from the notion of Gaia, in which all nature is connected and must be protected, if Wilson's thesis of interconnectedness is to be

---

[55] The unique role of water flow across ecoregions is worth additional note. Just think of the 6,853 km long Nile River, which most associate with the Nile Delta, the flooded savanna ecoregion in Egypt, but which has its origins half a continent away and tributaries with their origins in many other ecoregions. Or, think of the Ganges, Brahmaputra, and Meghna rivers, whose headwaters originate in the glacial ecoregions of the Himalayas and nourish many ecoregions as they flow to the Sundarbans mangroves of Bangladesh, commonly known as the Bengal Delta. Anywhere along these complex yet singular hydrological features, upstream ecoregions that are degraded by human activity can have catastrophic impacts on the ecoregions that lie downstream.

given proper consideration, one cannot randomly protect one equally sized ecosystem over another and think that the practical goal of 50% is being achieved.

It is worth revisiting each of these three issues in some depth.

What Do We Mean by Half?

Although we may now understand what we mean by "protected," what do we mean by "Half," as advocated by E. O. Wilson and the varied assortment of conservation-minded groups that insist "Nature Needs Half"? Might I suggest that there are some issues of measurement that must be resolved before we can even consider whether setting aside 50% of the planet from human incursion is anything more than a bold and catchy call to action? Perhaps this question is best answered by a series of three thought experiments:

a. If a pristine ecoregion characterized by north-south land animal migrations is bisected by an impassable east-west transportation corridor that covers only 1% of the ecoregion's landmass, should the ecoregion be considered 99% protected?

b. Say an ecoregion had two major halves that each provided essential ecosystem goods and services to each other, and later, one of those halves (50%) became developed into urban impervious surfaces, but the other half of the ecoregion is left unmolested. Should this ecoregion be considered 50% protected?

c. If one of two ecoregions of equal size is developed into urban impervious surfaces, but the remaining pristine ecoregion produces far fewer ecosystem goods and services for the planet and adjacent ecoregions, should this be considered a 50% compromise?

From these intentionally stark and simple thought experiments, you can likely generate countless additional questions as to how one might meaningfully calculate the necessary 50%. To simply calculate the portion of Earth's historical wildernesses (in square kilometers of surface area) that are not obviously disturbed by built structures and impervious surfaces, roads and other linear transportation corridors, deforestation and other methods of ecosystem transmogrification, and the like, is obviously insufficient. Whether this method is applied to the planet's geography as a whole, or to each of the world's ecoregions and then aggregated, it is insufficient.

Beyond the three wrinkles highlighted by the thought experiments above, there is still the issue of waste. How does one measure the proportion of an ecoregion that is stunted by persistent waste? Is it the surface area affected by this waste? If the ecosystem suffers no other human incursions, do we simply discount the viability of that portion of the ecoregion?

Lastly, how do we quantify the oceans? As mentioned earlier, oceans not only constitute 71% of Earth's surface. They also contain 97% of Earth's water and 99% of the living space on Earth. And despite all of the great progress in delineating and measuring distinct terrestrial ecoregions, no analogous effort has been launched for the deep oceans, outside of coastal ecoregions.[56] Not only would such

---

[56] Although this book focuses primarily on terrestrial ecoregions, the same biogeographic methodology has been employed for maritime environments. In 2007, the WWF and the Nature Conservancy project on *Marine Ecoregions of the World: A Bioregionalization of Coastal and Shelf Areas* offered the first-ever comprehensive marine classification system with clearly defined boundaries and definitions, representing broad-scale patterns of species and communities in the ocean. This effort was designed as a tool for planning conservation, yet it focused only on coast and shelf areas, and it did not provide any discussion of pelagic (open ocean) and benthic (ocean floor) environments. This means that many of the parts of open ocean, from surface to floor, that host the migrations of sharks, whales, dolphins, seals, sea turtles, and the like are not yet mapped as coherent ecoregions that are considered deserving of equal protection.

ecoregions need to be conceptualized in terms of stratified volumes, they would also need to be understood in terms of their distinct ecological relationships with the landmasses that they abut, relationships due to the flow of freshwater sources or the continuous exchange of fresh and saltwater in complex coastal ecosystems.

For instance, as discussed previously, the Gulf of Mexico is strongly influenced by the highly compromised ecoregions of the American Midwest, which are streaming increasingly large volumes of surplus nutrients into the Mississippi River, helping feed the growth of a massive dead zone. Other rivers traverse the watersheds of terrestrial ecoregions in order to feed freshwater into brackish ecosystems, such as the Chesapeake Bay, where unique species spawn. River deltas, tidal marshes, and estuarine waters, such as those of the Niger River Delta, are often extremely lush and fertile ecoregions that are continuously recharged by the nutrient-rich waters of ocean tides. We must not forget that often the ecological exchange between ecoregions is as important as the internal coherence of the ecoregion itself. And too often, these geographically narrow slivers of land (whether deltas, marshes, estuaries, mangroves, swamps, or what have you) at these exchange boundaries are overlooked, although they actually play a disproportionately important ecological role.

These and other complexities make it not so simple for us to calculate whether our planet's ecoregions are more than 50% protected or to determine whether this guideline meets the spirit of the Half-Earth concept, or even whether this benchmark, however calculated, meets the actual long-term ecological needs of our planet and our species.[57]

---

[57] Despite the complexities outlined here, the seminal work of Eric Dinerstein et al. in their 2017 *BioScience* article entitled "An Ecoregion-Based Approach to Protecting Half the Terrestrial Realm" still gives us a rough-hewn understanding of the relative status of ecoregions with regard to the impact of humanity:

How to Think About Ecological History

When calculating the proportion of our planet's historical habitats that still exist, it is essential that we operate from a useful historical baseline. Again, as a historically stable concept, a focus on ecoregions is helpful. It is too easy to fall into the conceptual trap of taking today's ecological boundaries as given, when too often the past few centuries (or even just the past few decades) of human activities have radically altered them. Changing habitats must be viewed in terms of their historical boundaries within an ecoregion. Such an analysis highlights the ecological deficits accrued in specific geographies.

Although the ecoregions construct makes it easy do some basic geographic subtraction, deducting the areas affected by humans from the ecoregion boundaries, understanding exactly what ecological processes were present across these historical wildernesses can be a challenge. Depending on the era in which the ecological damage took place, this exercise can be easier or harder. In the 21st-century reality of high-resolution, space-based remote sensing, one can observe change to ecological boundaries with some temporal regularity and even determine the source of the ecological damage. This is the kind of activity undertaken

---

The largest category is Nature Could Reach Half, with 313 ecoregions (37%), followed by the 228 ecoregions classified as Nature Could Recover (27%). Half Protected remains a reasonable goal in these regions. Within Nature Could Reach Half, 119 (38%) ecoregions have greater than 20% of their land area protected; the remaining 194 ecoregions (62%) have limited coverage of protected areas but retain considerable intact natural habitat. To achieve Half Protected, these 313 regions require only an expansion of their protected area network. The remaining 207 ecoregions (24%) classified as Nature Imperiled have little natural habitat and will require intensive efforts to achieve Half Protected or even to conserve the fraction that remains.

It is especially striking that no ecoregions are classified as free from human footprint and that there is not even a "75% Protected" category. This should be startling and deeply unsettling to any reader. Most important to understand is that this measurement does not take into account the expansive geography of human waste discussed in earlier chapters. If it did, these estimates would be much lower.

by the likes of the World Resource Institute's Global Forest Watch initiative. Such high-resolution imagery data are available going back at least two decades. Beginning in 1974, the availability of lower-resolution Landsat satellite imagery made it possible to determine ecological incursions and boundary changes. Prior to 1974, all the way back to about 1920 or so, there are airborne photography surveys of many places—particularly of cities and regions on the cusp of large-scale infrastructure investment, which often required such surveys. Archives of ground photos, if the locations can be verified, can be useful evidence when reconstructing timelines of our ecological past back into the 1800s. Before the dawn of photography, our collective memory of various landscapes is driven by hand drawings, textual recollections, and ultimately, archeological evidence.

As suggested earlier, it may very well be a valid intellectual exercise to look at a place like the Sahara Desert, and to understand the lush, if perhaps fragile, green ecosystem that existed there before modern humans crushed it with the weight of their demands. Understanding what the historical ecosystem once was may inform ecological restoration investments that we think could have comparatively large payoffs toward the generation of ecological goods and services that some notional 50% scheme might require. However, this would require major archeological efforts with unclear returns on investment. Or, we may determine that only certain ecological devastation from the past 200 years is worth restoring—an exercise that could be informed with some modest historical inquiry and some rudimentary landscape analysis. But certainly, there is no excuse for not using satellite imagery from the past four decades to establish a meaningful historical baseline for ecological boundaries and levels of landscape industrialization.

How we think about ecological history is key to how we calculate the carrying capacity of our planet. And at its core, this is a geographical exercise. Ahistorical and ageographical thinking, in this day and age, is simply inexcusable.

The Challenge of Interconnectedness

Not all ecoregions are equal. They are not equal in terms of the amount of geographic area that they cover, or the types, volumes, and values of the ecosystem goods and services that they produce. It is not that they range on a spectrum of superior to inferior. It is simply that they are each very different in the value that they provide to the planet as a whole, and to other ecoregions to which they are connected, either by geographic adjacency or by some more distant connection.

As such, before the Half-Earth thesis can truly be met, much more progress must be made on how we account for the ecosystem goods and services that given ecoregions generate. We must also make progress on how we map the flows of these ecosystem goods and services and the dependencies that other ecoregions, both adjacent and distant, have on them. This study of interconnectedness and a rigorous spatial accounting could finally provide humanity with the basis it needs to conduct meaningful ecological assessments as we conceive of and before we green-light new development projects.

The challenge of ecological interconnectedness is central to truly understanding our planet. Yet this scientific frontier will take quite a while to develop and to inform policymakers. This is why it is useful to understand the current state of research into ecosystem goods and services, and how this research is informing policy making today.

Valuing Ecosystem Goods and Services

The term "ecosystem goods and services" or "ecological goods and services" has been used liberally in the preceding chapters. Although the term is rather self-explanatory, perhaps it is worth discussing in more detail. Ecosystem goods and services are the benefits that flow from the ecological functions of ecosystems. These benefits accrue to all living organisms—including plants and animals and the ecosystems that support them—not just to humans. When the term is discussed, some will focus on their importance to humanity's health, social, cultural, and economic needs. Others will focus on how these ecosystem goods and services contribute to the basic carrying capacity of Earth for all its species, including humans. Unfortunately, the various public bureaucracies that have grappled with how to measure and map these ecosystem goods and services each have different missions and priorities. Academics have also applied different disciplinary perspectives to the task, measuring and mapping these goods and services in different ways. As such, any discussion I might provide here could only be illustrative. Nonetheless, I will endeavor to make it a useful discussion.

The natural world provides us with essential services that generate the essential goods that we require for life. In the most simple terms, examples of ecosystem goods are clean air and fresh water. Examples of ecosystem services are processes such as the purification of air and water. The two are obviously related since ecosystem goods are the products of the processes and interactions of natural systems that we call ecosystem services. But there are many other examples of both ecological goods and the ecosystem services that generate them.

Food is an ecosystem good. It is generated by the ecosystem process of plants capturing energy from the sun and using photosynthesis to combine water and nutrients from the soil and carbon dioxide from the atmosphere, creating food for animals, including humans, which in turn reproduce and generate more animals. Animal reproduction is another ecosystem service. In turn, within a given food chain, some animals consume others as food.

Soil is an ecosystem good. It is generated by the ecosystem service of plant material decomposition. Soil is required for plants to grow, to grow food and other ecosystem goods that flow from plants such as fiber, timber, and medicines based on the genetic diversity within our soils.

Before photosynthesis, as an ecosystem service, can lead to the production of food, as an ecosystem good, seeds must be dispersed and pollination must occur. These processes are ecosystem services that are supported by bats, birds, insects, and other animals. The nutrients we derive from the plants that we consume, and the animals that consume them, are ecological goods resulting from many ecological services.

While ecosystem goods and services are valuable economically and can be measured in terms of dollars, many are also essential to our survival and to the proper functioning of ecosystems that sustain humanity. Complex combinations of ecosystem goods and services help maintain biodiversity and the resulting genetic diversity that provides resilience in the face of disease and blight. They are key to the decomposition of wastes and the filtering (detoxification) of various contaminants from water and air. As mentioned earlier, they are essential to soil and vegetation generation and renewal. Many ecosystem goods and services help control agricultural pests and other forms of pestilence. Wetland goods and services aid in groundwater recharge, which provides the pure water that plants, animals, and humans require. Wetlands also often aid in the mitigation of floods and droughts, the moderation of temperature extremes, and the force of winds and waves. Oceans, which constitute the vast majority of our planet's surface and volume of its ecosystems, generate all manner of goods. Perhaps most notably, oceans' complex ecosystem services generate what some estimate to be as much as 80% of the oxygen our planet needs to sustain life, all while absorbing enormous amounts of poisonous carbon dioxide that would harm many living creatures. This oxygen production is just one of many ecosystem goods and services that we require for greenhouse gas mitigation.

Some of these are what an economist might call a "public good," in that they contribute to a global commons from which no one can be excluded (i.e., they are non-excludable) and the use of these goods does not reduce their availability to others (i.e., they are non-rivalrous). A key feature of such public goods is that they cannot be withheld from people who do not contribute to them. In the world of ecosystem goods and services, in a way, we are all "free riders." We enjoy the benefits of the clean water and clear air that ecosystems generate even as we each inject waste and contaminants into our soils, surface water, and atmosphere. We enjoy the food that the planet helps generate, such as seafood, even as we each contribute to the continent-size garbage patches in our oceans. We enjoy the medicinal innovations borne of genetic diversity harvested from our environment while our industrialization of the global landscape has led to a vast genetic winnowing as it fuels a massive species collapse.

It is easy to be overwhelmed by the number and variety of ecosystem goods and services, and the complex interconnectedness between them. However, what is important is that we are losing these goods and services at an alarming and unsustainable rate, due to the industrialization of the global landscape, as land is converted from its natural state and loses its ecological functionality. Little has been done to rigorously account for the ecosystem goods and services that are generated and the demand that humans make on them. Fundamentally, the sustainability of human communities and economies depends on an ecological balance that offers just the right amount of ecosystem goods and services. Anything approximating such a balance has been difficult to imagine as the world population has more than doubled in the past half-century, and in turn we have managed to industrialize a much larger proportion of the global landscape, which used to generate all manner of ecosystem goods and services.

For the past several decades, after the dawn of the environmental movement, some sort of environmental impact study has often been required of new development projects. Such studies are essential because they can help determine the impact of human activities on the ecological landscape before those activities cause detrimental changes in the environment, to include changes to energy flow, water, and chemical balances. However, such assessments tend to be local in their focus and are always conducted in an ahistorical and ageographical manner. If the marginal ecological impact of each project had to be placed in the larger historical context of the broader regional geography, the analysts, citizen stakeholders, and policymakers would be kept informed about the stark ecological debit being deducted by each subsequent development. Without such context, each additional development project can be made to appear reasonable and sustainable, on its face.

When we began to industrialize the global landscape in earnest two centuries ago, we demonstrated little regard for the environment. As was pointed out previously, the concept of nature was still in its infancy and not broadly understood. Even with the environmental awakening that has steadily taken hold over the past half-century, the rate of landscape change has accelerated to accommodate the doubling of world population that has occurred during this time period. This change has carried an incredibly high cost in terms of water and air pollution, loss of natural areas and biodiversity to agriculture and (sub)urbanization, generation of epic volumes of persistent wastes across our global landscape, and precipitous loss of essential ecosystem goods and services that we need to survive. In doing so, these costs are being passed on to future generations in the form of an "ecological debt."

Thinking Like Aldo Leopold

The question now is whether our world's population and our demands for development can be put back in balance with the ecosystem goods and services that we need to not only survive, but ideally to thrive. First, this ecological debt will have to be paid down by reclaiming large chunks of the planet, from both its industrialized surface and its accumulated persistent wastes, in order to restore large volumes of ecosystem goods and services. Second, the human population will have to decrease, and its ecological footprint will have to lighten until there are sufficient ecosystem goods and services to support the population without incurring massive ecological debt. Third, the geographic distribution of humanity (what some call our planet's "human geography") may very well have to change, substantially.

Navigating our way to a lower, more sustainable population plateau will require us to think not only like Roosevelt, Pinchot, and Muir. It will require us to think like Aldo Leopold, with a mind focused on the restoration of many of our historical wildernesses. Just as Leopold and his University of Wisconsin colleagues gazed upon the ruined agricultural landscape of the Great Depression and were inspired to think of how they might restore the historical wilderness, we should consider ourselves challenged to look upon the vast landscapes that modern industrialized humans have decimated and think about how we might aid nature in its inexorable efforts to reclaim them. Nature has not surrendered the natural processes that would no doubt reclaim much that has been taken. The only reason it has not reclaimed its loss is humanity's relentless machining and tooling of Earth's surface, holding nature at bay while continuously seizing more of our planet from wilderness.

If Aldo Leopold were alive, no doubt he would encourage a deliberate, strategic effort to identify the critical ecological resources that have been lost to mankind, in order to facilitate these natural reclamation processes. He would recognize the need for efficient and sustainable human habitats to be carved out, and he would seek to sustain them and connect them by ecologically light forms of infrastructure. When Leopold died in in 1949, the world had roughly 3 billion humans. Did he anticipate the absurd population growth that would occur in the latter half of the 20th century? The ecological destruction? The persistent wastes? How would Leopold have tackled our current predicament? We will never know. But his restoration worldview should certainly inform any effort we may undertake to recover ecologically if we manage to find our way to a lower, more sustainable population plateau.

But before we could make determinations about how we might deploy various conservation-preservation-restoration strategies, we must understand the nature of humanity's connectivity, how that shapes our planet's geography, and the opportunities it might afford us to restore historical wildernesses as humanity continues to urbanize.

CHAPTER SEVEN

Connectography as Possibility and Peril

The global strategist, author, and world traveller Parag Khanna coined the neologism "connectography" to address how the world's geography is being reshaped by various forms of connectivity. His book *CONNECTOGRAPHY: Mapping the Future of Global Civilization* is a tour de force that explores how transportation, energy, and communication infrastructure, in particular, have conspired to build a global network of urban centers that will play a more significant role in shaping our planet than will political borders.[58] In his worldview, railroads, pipelines, and communication fiber in many ways constitute more significant lines on our global map than do many international borders. They facilitate the flow of goods, people, and services in ways that make the connected world more prosperous, and ultimately more stable and resilient.

Surely, modern forms of connectivity offer unprecedented possibilities for our global society, and the various nodes of civilization that are connected through this network. Commerce and cultural engagement are

---

[58] The future map of the world that Khanna presents is truly frightening from an ecological perspective. When overlaid on the global ecoregions map, the massive sprawling megaregions that he predicts fundamentally eliminate all ecosystem goods and services that historical wildernesses once offered our planet and our species.

key factors that actively militate against the sources of instability that seem too difficult to thwart through policy. After all, those countries engaged in trade are least likely to engage in military conflict.

And while it is not the main point of his book, Khanna also kicks off a valuable macro-level conversation about how we need a much more sustainable approach to the new infrastructure our global society is building in our quest for connectivity. He is not shy about pointing out the ways in which infrastructure and the global supply chains they enable have made it easier to rape our planet, and how the topographical engineering techniques that have been honed over recent centuries have cost us ecologically.

Yet it is not until his penultimate chapter that Khanna really shines a light on the ecological impacts that these nodes and corridors of connectivity have had and will have in the future. These networks across the landscape have, after all, grown from thin lines to massive infrastructure corridors that transect ecosystems in fundamental ways.

How the History of Connectivity Reshaped Our Planet's Ecological Future

While Khanna's treatise draws on the past, it is focused on the future—specifically, on how connectivity is geographically shaping the future of global civilization. Thus, Khanna spends less time discussing the evolution of humanity's connectography and the growth of connections from their earliest phases. Early roads and railways were just thin lines on the landscape, with limited geographic reach, and little impact on the world's various ecologies. Migrating animals could once cross those roads and rails with little to no impedance, and their reach was far from ubiquitous. But over time, these thin lines transformed into highly trafficked, impassable corridors that disrupted ecoregions all over the planet. Beginning somewhere in the middle of the 20th century, the nature of connectivity between cities and regions began to transform. Such transportation corridors of connectivity came to disrupt how the ecosystems that they transected functioned.

However, this connectivity went far beyond transportation corridors. The telegraph soon accompanied railways, bringing the first communication networks between cities. Wireline communication required electricity, first electrostatic charges and later running voltage. Soon electrical power lines joined these corridors, and then spread independently to bring power from power stations to anyone and any place it could reach. As the voltage increased, more and bigger power lines required a wide berth from nature in order to prevent fires and damaging encroachment from fallen trees. The resulting power-line corridors denuded large swaths of land stretching hundreds of thousands of miles. As oil extraction took hold in the late 19th century, so did oil pipelines, the first debuting in 1879. Soon these pipelines crossed entire continents. Power lines, communication lines, pipelines, and the like all came to demand the clear-cutting of corridors across vast landscapes, and continuous efforts to prevent nature or mankind from encroaching. Not only did these corridors fundamentally remake how humans experienced and perceived space and time in the movement of people, energy, commodities, and information, they also reshaped the basic boundaries and flows of many ecosystems.

It is important to note that connectivity's ecological impacts were never constrained to the land. Many of the regional and global networks connecting the world's centers of power have been over navigable waterways, whether inland, riverine, or maritime. Although humans have been using boats for millennia, the scale of maritime connectivity exploded over the last two centuries and its growth over the past half-century has been unrestrained. This process led to fuel spills, coal and then diesel emissions, anthrophonic pollution, and many forms of water pollution (from sewage dumping to bilge waste to garbage dumping), to the massive global spread of invasive species.

Commercial air travel is not even a century old. Powered, fixed-wing flight, achieved by Wilbur and Orville Wright in 1903, quickly evolved into commercial postal and freight carrying, and then quickly into passenger airlines. But it was the British de Havilland Comet, which entered service as the first purpose-built jetliner in 1952, that began more than a half-century of exponential growth in the number of flights connecting cities, countries, and regions. Remote air strips that had been established in the first half of the 20th century became massive international airports before the century was out and now move more than 3 billion passengers around the world every year. More people fly every year now than inhabited Earth in 1950. Contagious diseases can (and do) now travel from one continent to another in less than a day.

Transportation lines, lines of communication, pipelines, and power lines, spanning land, air, and sea, have driven economic growth and rising standards of living through trade, social cohesion through cultural exchange, and resiliency through rapid response to emergent situations. Their ecological impact has been equally dramatic. And even as communication turned wireless with the dawn of radio, television, and cellular infrastructure, without the need for massive destruction of ecological corridors, this connectivity quickly drove a global social awakening that has created an insatiable demand by the peoples of the world for the living standards of the developed world. Ultimately, this insatiable demand, as manifested over the 20th century, led to much more ecological destruction than the transport and communication corridors themselves.

Global Supply Chains and Their Ecological Footprint

In important ways, all of this connectivity has evolved as part of the global supply chains that have enabled humanity to harvest the bounty of our planet (or use the planet to generate all manner of unnatural bounty), to add value through processing and production, and to move these resources efficiently to wherever global demand dictates. Khanna refers to these ecological resources as "natural infrastructure" that we have spent the past decades, centuries, and millennia building physical infrastructure to exploit. These global supply chains have enabled humanity to haphazardly erase wildernesses and ecological resources in distant lands, if doing so happens to feed the end goal of meeting global demand most efficiently or costlessly. With few exceptions, globalization has not yet demonstrated an ability to treat these natural resources as strategic resources in the achievement of some better balance between man and nature.

These global supply chains each have ecological footprints in terms of the portion of Earth that they have industrialized, the resources that they consume, the industrial wastes that they generate, the environmental costs of various forms of connectivity (whether for transport, energy, communication, or other purposes), and the postconsumer wastes that we accumulate as a global economy that is premised upon planned obsolescence and disposable consumerism.

Khanna is right to point out the ways in which the ecological footprints of the global supply chains can gather in places that are distant from the actual demand that drives the industrial production at its core. As he points out, some 40% of "Chinese" emissions are attributable to Western companies that have outsourced their manufacturing to China due to its cheaper labor and often more lax environmental regulations. The same could be said for so many of the agricultural items to which developed nations have become addicted. Global supply chains have enabled developed nations to never have to forego their favorite foods simply because they are locally out of season. Instead, distant nations have obliterated historical wildernesses in order to supply tropical fruits, vegetables, fish and shellfish, and the like all year long to massive, voracious populations in developed nations that give those nations economic incentive to continuously expand agricultural production. This may be most acutely felt for crops such as palm oil and corn, where the transformation of agribusiness into an insatiable industrial processor of agricultural raw materials has led to the replacement of vast wildernesses with monoculture crops.

Khanna is also right to point out the burdens on distant water resources that these global supply chains have created. Of course, the consumers of end products are rarely aware of the water crises that their demand is creating at the far end of these global supply chains. Demand at one end of the planet can induce social and political crises at the other end as precious water resources are diverted from ecosystems and human communities to support the industrial production that supplies this global demand.

As we will see in the next chapter, there are reliable ways to calculate global ecological footprint issues, if only by national scale. Whether your nation is a net ecological debtor or creditor depends on how it participates in these global supply chains.

When Nature Has Its Way

Still, Khanna seems to frame many of these issues of globalization, urbanization, and infrastructure connectivity in terms of human conquest, the revenge nature might exact, and how humans can adapt. Although he calls for more strategic action in pursuit of a better balance between the needs of humankind and nature, his chapter "When Nature Has Its Way, Get Out of the Way" still frames environmental considerations as something that an ever-growing human population can address through more sustainable investment and adaptations.

He addresses head-on the impending threat of sea-level rise that will require a retreat from the water's edge. The impact this threat will have on hundreds of our planet's largest cities, not to mention our island nations, informs Khanna's sobering discussion of how we will need to think strategically if we are to accommodate the billions who will likely become climate refugees by 2100, if not sooner. He even references the very real likelihood of ecological collapse that various regions will face, whether due to the depletion of water resources or to some other cause.

Yet his project focuses more on the opportunities for engineering solutions to these ecological challenges, whether they be geo-engineering solutions or more of the topographical engineering that has underpinned humanity's mastery of the wildernesses on which we have built our modern industrial civilization. Khanna warns that humanity can steer nature, but not fully control it, and asks us to think about how we can better negotiate with nature as it challenges us with floods, droughts, chronic water shortages, food crises, sea-level rise, and all manner of natural disasters.

To be fair, Khanna's project is not intended to consider the fundamental carrying capacity of the planet, like this book and those thinkers whose theses are addressed in the next chapter. He does not begin from a consideration of Earth's "people problem." In a way, he assumes that population curves will simply go where they go, and humanity will have to adapt by means of urbanization and connectivity through sustainable infrastructure. Fair enough. Nevertheless, Khanna sets the stage for us to explore the geostrategic importance of considering the ecological opportunities inherent in our investments in connectivity.

The Ecological Opportunity in Connectivity

Connectivity, and the connectography that it induces, also present enormous opportunities to limit humanity's ecological impact on our planet. A global network of sustainably engineered cities that are connected in ways that do not haplessly transect vital ecosystems would be far superior, ecologically, to the endless sprawl of low-density suburbs, exurbs, and informal settlements fed by massive networks of surface roads that have transformed millions of square kilometers into highly trafficked impervious surfaces that aggressively repel the historical natural processes of the region that they cover.

Underground connectivity, whether for transportation, power transmission, or wireline communication, can avoid all manner of ecological disturbance, limiting the need for the systematic denuding of terrestrial corridors. Elevated infrastructure has the potential to lighten the terrestrial footprint of various forms of infrastructure, though most engineering approaches demand denuding the surrounding land in order to avoid damage due to encroachment by nature. Wireless connectivity and various forms of airborne connectivity have similar ecological benefits, by reducing the terrestrial footprint. However, aircraft and the aerospace industry have a long way to go to mitigate the wide variety of ecological impacts that they precipitate due to sound, emissions, and endocrine-disrupting effluents.

Each new generation of engineers learns from the mistakes of those who preceded them. And the lessons they learn reinspire the core design and engineering principles that become commonplace across their profession. Much of the trend toward sustainable engineering techniques has been in response to the heavy-handed and highly destructive engineering methods of the 20th century, both in the private sector and by public sector engineering institutions such as the U.S. Army Corps of Engineers. Yet the ecological footprint of some forms of connectivity has not yet been really re-examined and re-envisioned by the global community of engineers. Can more energy-efficient, less ecologically impactful forms of mobility be developed that do not demand millions of square kilometers of pavement and impervious surface? Can electricity generation and distribution be engineered to be more sustainable and more efficient, and to not require the removal of millions of linear miles of trees? And can legacy productive endeavors (farming, mining, manufacturing, etc.) be re-engineered to have less intensive land use demands and be arrayed across these other connectivity networks to ensure more sustainable and efficient production and logistics?

There is great possibility in deliberately engineering our future world's connectivity to achieve a fundamentally new, ecologically minded geography for humanity. This future connectography, if thoughtfully conceived, could not only drive a more politically, socially, and economically resilient civilization. It could also help humanity thrive while thoughtfully and strategically minimizing its ecological footprint. As Khanna points out, this more sustainable approach to urbanization and connectivity could help kick off a process of "giving back" to nature, as humanity abandons former human habitats and allows a resilient Mother Nature to gradually reclaim them back into her historical ecosystems.

The Peril Is Real

The peril we face, however, is profound. If we continue to industrialize the planet's surface through the same heavy-handed engineering of ecologically destructive forms of connectivity, particularly when designed to enable widely distributed forms of land use, we could very well "pave paradise," as the song goes. Geography and the topographical engineering approaches that it co-evolved with in the 18th, 19th, and 20th centuries could easily continue to be used to enable humanity's "conditional conquest" of our planet. If we travel this route over the next few decades, there is little hope for our species as we will undermine the planet that sustains us.

Geography, however, could actually be the source of our salvation, rather than a tool of our destruction. As Khanna says, if we map our critical ecological resources and fully label these resources on our maps, we could better manage them, and perhaps even better manage the balance between nature and mankind. Why should our maps hold only names of political units and boundaries, our mountain ranges, and our oceans and seas? Khanna rightly asks why the rest of the map should simply be color coded green for forests, tan for deserts, brown for mountains, white for ice, and blue for water.

Khanna is clear that there are sacred ecological geographies at the core of not just our global ecosystem, but the complex global system at the heart of our very human geostrategic considerations. Perhaps if we elevated these critical ecological features in our cartographic portrayals of Earth in the maps and globes used by everyday citizens, business leaders, and public sector decision makers, we might find ourselves collectively working just as hard to protect nature's boundaries as we do to defend our political boundaries. As a global society, we would then be able to shift from a myopic dependence on political geography to being empowered by a functional geography that understands the nuances of how we, together, can mold our human geography to complement the biogeography of our finite and singular planet.

Until we do that, we are all in peril.

CHAPTER EIGHT

How to Calculate Earth's Maximum Carrying Capacity

To be blunt, humanity's recurring and accumulated wastes—by themselves—are on track to squeeze out the ecological resources required to support our species over the course of this century. Our industrialization of Planet Earth not only has eliminated a certain portion of the ecological carrying capacity of the planet. Our mode of industrialization also is designed (if inadvertently) to regularly and reliably generate increasingly more wastes proportionate to the size of the population that inhabits this landscape.

Modern industrialized humanity does not live in the understory of forest wildernesses or live in sustainable ecological partnership within various other ecosystems. Modern industrialized humans remove these ecosystem resources, and actively machine and tool the geographies that we inhabit and cultivate, in which we recreate, and across which we transport people and goods. We do so to meet our needs, which include ensuring that nature is prevented from reclaiming its historical ecological boundaries. A future population of any more than 3 billion humans would continue to accrue additional ecological debt by continuing to machine, tool, and lay waste to ecologically critical portions of Earth's surface and to spoil ecologically critical portions of Earth's oceans. And the human population is projected to continue growing exponentially for some time.

World Population, 2010 and 2030

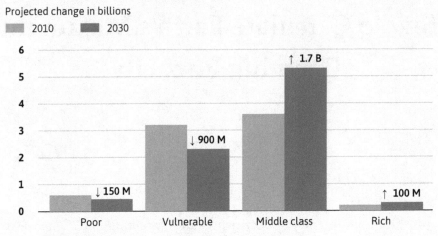

Source: World Data Lab

In 1950, when the population of Earth stood at about 3 billion, a large proportion of America's historical ecosystems had been annihilated in order to support farming, grazing, logging, mining, transportation corridors, urbanization, suburban housing, and the like. It is hard to calculate a similar metric in 1950 for other developed nations since much of Central Europe, Eastern Europe, Russia, China, and Japan had been ecologically compromised by years of total war. Around 1950, many lesser-developed geographies still had modest populations and little in the way of ecological incursions by their populations. But the population curve was clear. Nations around the world, of all development levels, were experiencing their own "baby booms." Now, only seven decades later, we stand at just over 7.5 billion.

As noted earlier, growth forecasts put the human population at 9 billion by 2050. Without a rapid subsequent contraction in numbers and rewilding of our ecological resources, this population will crush our planet's long-term carrying capacity, and the fuse will be lit. Even if we spent the next several generations seeking to squeeze out nearly all waste by-products and inefficiencies from the modern human modus operandi, and began rolling back the geographic footprint of humanity considerably through densification strategies, these current population projections spell doom.

Remember the 200,000 years it took to get to 3 billion humans. More sustainable agricultural and manufacturing practices than we used in the 1950s may now be in use in many places across the American and global landscapes. However, the sheer quantity of human demand for goods and services has exploded, with per capita annual economic output in the United States alone increasing nearly fourfold since 1950, and the addition to the global middle class of another couple billion people around the world with similar consumption patterns.[59] If the entire global population were already in the global middle class, under our current "production function," its consumption would have already obliterated our fisheries, ocean dead zones would already have crushed ocean species and the oceans' ability to generate oxygen, deforestation in support of growing crops would have already crippled more critical ecosystems, and our many species that are currently at the brink of collapse would already be long gone. Certainly, even if this century spawns an unprecedented wave of sustainable technological innovations that help reduce the median human ecological footprint, the notion of a world population of 9 billion or 11 billion is ecological folly.

---

[59] By the calculations of the World Data Lab (www.worlddata.io), as of 2018 a majority of the world's population can be classified as either rich or middle class. Although this "middle class" category is broad and enjoys lifestyles that differ widely, some 3.59 billion people now occupy this category, with another 200 million people categorized as rich. However you count it, we are talking of at least a couple billion people living highly consumptive lifestyles that have second- and third-order effects on our planet's ecological wellbeing of which few consumers are conscious. It is projected that by 2030, this middle class will include some 5.3 billion humans, with 300 million categorized as rich. One can also be confident that the consumption patterns of the middle class of 2030 will be substantially more burdensome on our planet.

Again, this is not a Malthusian argument. I have no doubt that capitalism could generate enough food and water to keep 9 billion or 11 billion people served well above their minimum caloric requirements, protein demands, and freshwater needs. The question is at what costs. At what ecological costs? At what geopolitical costs? At what moral costs? Would these structural costs incur a long-term ecological debt? And ultimately, how long until this ecological debt comes due? Some scientists calculate that the sixth mass extinction could culminate by 2100, if the human population and our behaviors do not change radically quite soon, due to the vast array of human ecological impacts.[60]

It is my estimate, as a technological optimist, that Earth can support no more than 3 billion modern industrialized humans over the long run, if we seek a human species that can thrive—not suffer unspeakable deprivations on the journey to planetary annihilation. So, how did I arrive at this estimate of Earth's maximum carrying capacity for human life? In truth, there are several well-established lenses through which we might assess the issues at hand. Despite being well accepted, I would argue that some of these approaches to calculating Earth's carrying capacity leave us ill equipped to deal with the challenges that we face. At this point in the discussion, you will not be surprised to learn that I favor

---

[60] The WWF's biennial *The Living Planet Report* details the state of the planet and its implications for humans and wildlife. The 2016 report found that the number of vertebrates in the world fell by well over half between 1970 and 2012 and that, without intervention, the decline will continue, projecting that 67% of all animals will be gone by 2020. There is no doubt that the factors called out in the report will precipitate a wave of extinctions that will unfold at an unprecedented pace. Many destructive forces lead to species extinction. Some scientists, such as Daniel Rothman, a geophysics professor at MIT, focus solely on the scourge of atmospheric carbon which, one way or another, was the common culprit in the previous five major extinctions. His analysis, which took into account 31 major changes in the carbon cycle over the last 542 million years, indicates that human disruptions to the carbon cycle over the past two centuries alone are on track to drive a sixth extinction, given the historical impacts of surpassing various carbon thresholds.

methodologies that are grounded geographically. After all, if we cannot account for the ecological carrying capacity of our finite planet geographically, what do we think we are doing?

Perhaps it is useful to begin with one of the more prominent assessments of the late 20th century that hinges on major assumptions about the energy requirements of modern humans. We will start with the considerations and calculations of Anne and Paul Ehrlich, authors of *The Population Bomb* and *The Population Explosion*,[61] as further addressed by former Presidential Science Advisor John Holdren.

---

[61] Stanford ecologist Paul Ehrlich's 1968 book *The Population Bomb* was the first to trace our planet's onslaught of ecological destruction to "too many people." His book rode the crest of popular concern about the environmental implications of the post-World War II global population boom. This broad-based concern led President Nixon and the U.S. Congress to create the Commission on Population Growth and the American Future in 1969. Alas, such talk was ultimately sidelined as alarmist, and discussions of the "Limits to Growth"—spurred by the 1972 eponymous computer simulation study commissioned by the Club of Rome—were dismissed by those who argued that the study's methodology was flawed. The study found that limits to growth on Earth would become evident by 2072, leading to "sudden and uncontrollable decline in both population and industrial capacity." Criticism even came from the premier growth economist (and Nobel Laureate in 1987) Robert Solow from MIT, who argued that the predictions were based on weak data foundations. In essence, critics argued that the model failed to allow technologies for expanding resources and controlling pollution to grow to keep up with the population, capital, and pollution. Some forget the role of Julian Simon in this debate. As detailed in the 2013 book *The Bet: Paul Ehrlich, Julian Simon, and Our Gamble over Earth's Future*, by Paul Sabin, Simon played antagonist to those concerned about population, the environment, and limits to growth—including Ehrlich and his 1971 "Impact of Population Growth" co-author John Holdren (later President Obama's Science Advisor). Simon, a conservative economist from the University of Chicago, argued that markets would allocate scarce resources and stimulate innovation to solve any population pressures. These critics were technological optimists or simply individuals blinded by the assumptions of neoclassical economics. *The Population Explosion*, published jointly by Paul and Anne Ehrlich in 1990, sought to refocus attention on the numbers, stating that "Then the fuse was burning: now the population bomb has detonated." Still, in all of these arguments, pro and con, there was no real discussion of how to calculate the long-term ecological carrying capacity of our planet when populated by modern industrial humans.

The Energy Thesis of Optimal Global Population

In his 1991 article entitled "Population and the Energy Problem," Holdren looked at how a growing population stressed the world's available energy supplies and then considered what would happen if everyone had equal access to energy.[62] This paper built upon an equation that Ehrlich and Holdren had developed in 1971 to assess human impact on the environment: I = PAT, or Impact = Population × Affluence (consumption per person) × Technology (damage per unit of consumption).

As a point of reference, global energy consumption in 1970 was 8.3 terawatts (TW) and had grown to 13.2 TW in 1990, at the time of Holdren's writing. Global energy consumption in 2018 was about 19 TW.

Holdren assumed an increase in the human population to 10 billion and stipulated a reduction in the average per capita energy use in industrialized nations from 7.5 to 3 kilowatts (kW) annually. (Remember, this 7.5 kW average was an average based on 1990 behaviors in the developed world.) In order to achieve global parity, he also stipulated that average per capita energy use in developing nations will increase from 1 to 3 kW. Note that at the time his article was published, this reduction would have demanded that citizens of the United States cut their average energy consumption from almost 12 kW to 3 kW, which it was assumed could be achieved through the adoption of existing energy-efficient technologies and some improvement to most people's standard of living.

From all of these assumptions, Holdren thus envisioned a world that would see total global energy consumption increase more than a third over

---

[62] This article explicitly built upon the formulas first presented by Ehrlich and Holdren in their 1971 article "Impact of Population Growth," with a particular focus on how limits to energy production and the ecological impacts of energy consumption presented real constraints on the size of the population that could live sustainably on Earth over the long term.

consumption today—to 30 TW, from the 3 kW of energy consumed by each of 10 billion people. The mix of energy sources that would be used to achieve this was unclear.

Two and a half decades later, current estimates expect global energy consumption to hit nearly 25 TW by 2040 and 28 TW by 2050 as the population grows and more people have access to energy-intensive lifestyles. And frankly, I strongly believe that these estimates fall far short, assuming vital energy efficiencies materialize while failing to understand the explosion of the global middle class and just how electrified all our futures will be.

In 1994, Gretchen Daily, Anne Ehrlich, and Paul Ehrlich postulated an "optimum human population size" on the basis of Holdren's paper. They based their calculations on the energy consumption split that they observed between rich and poor countries, at the time—70% rich, 30% poor. They proposed a world that relied on 6 TW of energy and a society that required only 2 kW per capita and concluded that an optimal global population would have to be 2 billion or less. "Optimal" is certainly a loaded term, and they used it to encompass all manner of value assumptions. But their mathematical exercise forced one to think about the orders of magnitude we are dealing with regarding such issues.

Changes in energy consumption in the past two decades raise other questions that this analysis glided over. Rather than 12 kW a year, Americans in 2018 consumed 13 kW on average. The global average per capita consumption of energy was 2.5 kW. Although the mix has changed, over 78% of this power generation globally was still from fossil fuels, meaning that the volume of emissions has only increased. All this while no end of new, energy-efficient technologies have become available to the average consumer. As for the split between the rich and poor countries, many countries have moved up the development curve over the past three decades, consuming energy in ways that more resemble the United States than the remaining poor countries.

This means that the 3 kW per capita that Holdren projected for a future that he postulated held 10 billion people was likely an unrealistic limit. The 2 kW per capita assumed by Daily, Ehrlich, and Ehrlich was even more unrealistic, particularly in a world that is now defined by so many humans holding personal supercomputers in their palm. As every house and every city becomes "smart," with the dawn of the Internet of Things, ubiquitous sensors, and everyday robotics, and as the mobility of the global population expands—and as these technologies became mainstream for the global middle class—per capita energy consumption will only increase.

In the 1990s, some believed that there was a looming threat of "peak oil" production that would ultimately constrain human energy consumption if alternative forms of energy did not take up the slack. Holdren, Daily, and the Ehrlichs all stipulated that per capita energy consumption must decline or the population must decrease, unless there is a massive growth in the use of renewable energy. They understood that allowing fossil fuel consumption to grow to serve a world population of 10 billion, with the majority of them having joined the global middle class, would be ecological suicide.

Perhaps in their own form of energy Malthusianism, they could not imagine a world where renewable energy production could sustain 10 billion at the energy consumption levels of the developed world. However, with current technological realities, it is clear that solar energy alone, paired with a range of energy storage solutions, could easily meet these energy needs in the coming decades. This is particularly clear as solar generation becomes increasingly geographically co-located with energy consumption, vastly reducing line loss. This does not even account for the wide variety of other renewables, and yes, Generation IV nuclear energy solutions that are on the horizon. One could even assume 10 billion people consuming five times the energy that the most profligate of us do today—yes, more than 100 TW—all fueled with renewable energy.

Energy is not the limitation. At this stage, meeting current energy needs would require only investment in highly lucrative, unsubsidized, renewable energy generation and storage assets. Daily and the Ehrlichs' "optimal population" of 2 billion, driven by their energy calculus, simply does not hold water because energy is not the limiting factor in this situation.

So, to understand whether Earth's population must be curbed, we must look somewhere else.

The Planetary Boundaries Thesis

Another well-established viewpoint on our planet's ecological limitations is the so-called "Planetary Boundaries Thesis." In most of its formulations, this thesis holds that there are clear thresholds which, if surpassed, will serve as a kind of ecological tipping point. This notion assumes a kind of non-linear dynamic in which a seemingly small marginal increase in, say, atmospheric carbon, would begin to produce large, possibly catastrophic change.[63]

The idea of a boundary was advanced to account for complexity and uncertainty, offering a safe range or distance from a given threshold. Stay below the boundary, and we will safely avoid the threshold. However, if we cross the boundary, we enter a danger zone where peril looms.

---

[63] This reference to the nonlinear nature of the dynamics highlighted by the Planetary Boundaries community is in no way meant as a criticism. If anything, I would argue that humanity is often caught off guard by so-called Black Swan events because we are conditioned to believe that trend lines will simply continue on as they were. Unexpected events of large magnitude and consequence, and their dominant role in history, should disabuse us of that notion. As Nassim Nicholas Taleb explains in his book *Black Swan: The Impact of the Highly Improbable*, such events play vastly larger roles in shaping our history than do regular occurrences. The Planetary Boundaries community is right to be focused on the thresholds that, if crossed even by only seemingly marginal amounts, will have devastating consequences. The threats we face are threats of discontinuous change, to which humanity, along with many other species of flora and fauna, will find it difficult if not impossible to adapt.

This thesis, advanced by Stockholm Research Center director Johan Rockström and a group of 28 internationally renowned scientists, was intended to move the discussion away from "limits to growth." Rather, they sought to move the discussion toward a governance and management approach to identifying an estimated safe space for human development, a space in which we could avoid global-scale, human-induced, irreversible environmental damage. In doing so, they seemingly opted out of the larger discussion of how the size of the human population affects our planet.

This team of scholars and practitioners, in a 2009 study, identified nine planetary boundaries.[64] They asserted that transgressing one or more of these boundaries could be highly damaging or even catastrophic to life on Earth, due to the risk of triggering non-linear, abrupt changes to planetary-scale systems. The team sought to estimate how much further the human ecological footprint could increase before "planetary habitability" would be threatened by climate change, biodiversity loss, and biogeochemical flow.

The nine boundaries that must not be crossed if humanity is to survive:

1. Climate change: Carbon dioxide concentration in the atmosphere of less than 350 ppm and/or a maximum change of +1 W/m2 (watt per square meter) in radiative forcing

---

[64] The Stockholm Research Center team identified the processes that regulate the stability and resilience of Earth as a system. The nine planetary boundaries were quantified, with the proposition that humanity can continue to thrive for generations to come, as long as these boundaries are not breached. If we were to cross these boundaries, humanity would face an increased risk of generating large-scale abrupt or irreversible environmental changes. The seminal article that launched this body of thought into the popular discourse was "Planetary Boundaries: Exploring the Safe Operating Space for Humanity," published in 2009 in *Ecology and Society*, with a 2010 TED Talk by Rockström driving popular attention to the concept.

2. Change in biosphere integrity: Extinction rate of fewer than 10 extinctions per million species per year

3. Stratospheric ozone depletion: Stratospheric ozone concentration maintained at less than 5% reduction from the pre-industrial level of 290 Dobson Units

4. Ocean acidification: Global mean surface-seawater saturation state with respect to aragonite (a form of calcium carbonate) equal to or greater than 80% of pre-industrial levels

5. Biogeochemical flows: Anthropogenic nitrogen removed from the atmosphere limited to industrial and agricultural fixation of 62 million metric tons per year and anthropogenic phosphorus going into the oceans not to exceed 11 million metric tons per year

6. Land system change: 75-54% of original forest cover, as an average of three individual biome boundaries: tropical: 85% (85-60%), temperate: 50% (50-30%), and boreal (85% (85-60%)

7. Freshwater use: Global human consumptive use of water of less than 4,000 km^3/yr (cubic kilometers per year) of runoff resources

8. Atmospheric aerosol loading: Overall particulate concentration in the atmosphere, on a regional basis, at an amount to be determined

9. Novel entities: Concentration of toxic substances, plastics, endocrine disruptors, heavy metals, and radioactive contamination into the environment at amounts to be determined

Of these nine boundaries, the researchers have now proposed quantifications for eight. Yet all of these measures were oddly devoid of any geographical anchoring, save for the most recent iteration of the "Land system change" measure. Perhaps this is due to the climate science orientation of so much of the work, treating everything as fluid, interconnected global resources. All of these measures seem to be devised in that similar amorphous manner, even though many of these boundaries can really be understood only in terms of the very specific geographies and ecologies that humanity has spoiled, undermined, and even eliminated from the functioning of a planet with very specific geographic interdependencies.

The "Land system change" boundary is particularly curious, as it has changed over time. Originally, it measured only the land surface converted to cropland, somehow overlooking the enormous swaths of land ecology that have been paved with impervious surfaces, whether roads, parking lots, or buildings. It also overlooked the wholly unnatural introduction of vast lawns. Although green and permeable, this human-induced landscape change has served to annihilate all manner of ecosystem resources while giving the illusion of "green space" that humans have too often looked at with ecological pride. In addition, it did not account for land degradation, which may not result in a change in land cover but does result in impaired ecosystem function. Since then, the measure has been altered to basically be a percentage of original forest land that has been removed—a peculiar measure that seems solely focused on the land's role as a carbon sink.

As we have seen from the previous line of argumentation, any discussion of ocean acidification, biogeochemical imbalances, biodiversity loss, freshwater diversion, and chemical pollution fundamentally requires a geographical lens. Without understanding these phenomena in terms of the specific biogeographical dynamics at play, as they evolve over time, so-called "planetary boundaries" scientists do these important issues a disservice.

In the end, while drawing attention to the limited biocapacity of our planet and the pressures humans put on it, the Planetary Boundaries community dodges the hard questions about population, as though technological innovation will allow a population with unfettered growth to live within the boundaries identified.

The Ecological Footprint Thesis

Another way of calculating the carrying capacity of our planet is decidedly more geographic in its approach, yielding a more tractable and meaningful framework for understanding how humanity has taxed our planet. This is the "Ecological Footprint" effort of Mathis Wackernagel and the Global Footprint Network, which derives each country's ecological footprint from the biologically productive area that it takes to absorb a population's carbon dioxide emissions and to generate all the resources that it consumes.[65] Although this framework is geographic only insofar as it is tied to the concrete geographies of the world's nations, it is a good start. After all, nation states are the sovereign entities that may have some hope of taking action to curb humanity's ecological footprint, one nation at a time.

In this model, a country's consumption is calculated by adding imports to and subtracting exports from its gross domestic production. It is assumed that all traded commodities carry with them an embedded amount of bioproductive land and sea area necessary to produce them and sequester the associated waste. Each country has a certain amount of biocapacity, calculated geographically, assuming current technology and management practices, which differ country by country, depending on the productivity of their ecosystems.

From this analysis, we are left with a rough-hewn geographical analysis of those countries that have an "ecological reserve" or an

---

[65] This idea was first advanced by William Rees, an ecologist at the University of British Columbia, and his student Mathis Wackernagel in the 1996 book *Our Ecological Footprint*.

Planetary Boundaries Thesis

■ Below boundary   ■ Beyond zone of uncertainty (high risk)

■ In zone of uncertainty (increasing risk)   ☐ Boundary not yet quantized

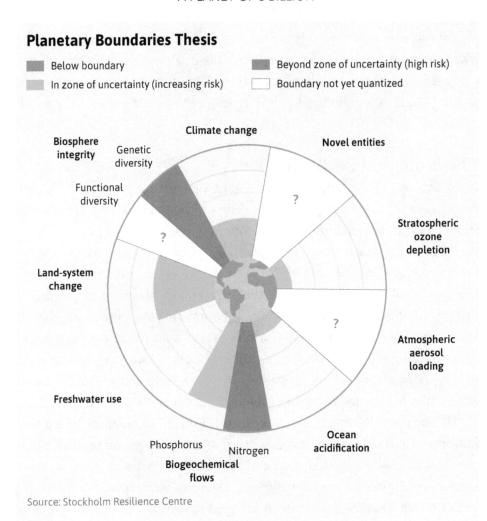

Source: Stockholm Resilience Centre

"ecological deficit," depending on whether their ecological footprint is smaller or greater than their biocapacity. These nations can also be thought of as "ecological creditors" or "ecological debtors" when putting them in the context of humanity's global ecological footprint. In this accounting, most of the major industrialized countries have long been running ecological deficits, fueling a global ecological deficit that this school of thought terms "ecological overshoot."

National geographic boundaries drive this analysis, contributing to the formulation of national footprint accounts that take into account measures of biocapacity, yield factors, equivalence factors, and specific land use types such as cropland, grazing land, fishing grounds, forest land, carbon uptake land, and built-up land. This analysis focuses heavily on the ecological impacts of various modes of agriculture, rather than, say, urbanized impervious surfaces. This is fair enough, given the vastness of agricultural development and the much larger proportion of Earth's surface that it commands.

The Global Hectare

Central to this method of geographically accounting for ecological carrying capacity and ecological burden is the notion of the "global hectare" (gha). This unit of measure is very powerful, albeit also glossing over much complexity. It quantifies the ecological footprint of people or activities, as well as the biocapacity of Earth, and when contrasted geographically using national boundaries, it quantifies the same for each nation. One gha represents the average productivity of all biologically productive areas (measured in hectares) on Earth in a given year. This measure enables the derivation of others such as "global hectare per person," as an average measure.[66]

The Global Footprint Network calculates that there are approximately 12 billion hectares of biologically productive land and water areas on Earth. It also calculates that humans are currently consuming roughly the equivalent of 18 billion hectares of biocapacity. Thus, humanity is consuming 1.5 planet's worth of biocapacity. In other words, our consumption is 50% too great for our planet. With approximately 7.5

---

[66] For quick reference, a hectare is defined as a metric unit of square measure, equal to 100 ares (10,000 square meters or 2.471 acres).

billion people on Earth, that comes to 1.7 gha per person. By this method's calculations, more than 80% of the world's population lives in countries that are running ecological deficits, where they use more ecological resources than their nation's ecosystems can renew. As of 2017, this approach calculates that the consumption of the current world population exceeds Earth's total ecological capacity by one half. When these measures are projected back in time, this calculation has humanity's ecological overshoot commencing in 1970. When exactly humanity began exceeding our planet's long-term carrying capacity is not quite the point of this chapter, but it certainly is a corollary of sorts. Above, I have asserted it was a bit earlier, around 1950. We will quibble about such calculations later.

If We Were All Swiss

This math can clearly be applied to the question of the carrying capacity of Earth. If we were to assume that Wackernagel's approach were correct in its estimate of the long-term ecological carrying capacity of the planet, and if we assumed today's mix of technologies, and if we were to allot each human on Earth the generous per capita consumption of ecological resources from Wackernagel's homeland of Switzerland (where people are hardly known for suffering!), we would be left to assume that Earth can only handle far fewer people than the 7.5 billion we currently have on our planet.

Using the latest data from 2016, the ecological footprint in Switzerland was 5.8 gha. The global biocapacity was 1.7 gha per person. If we divide Switzerland's ecological footprint by this per capita global biocapacity (5.8/1.7), we would require some 3.3 Earths for everyone to live like the Swiss. This means that under this methodology, Earth could support roughly 2.4 billion people. Obviously, to live like the average American would allow for far fewer people on Earth, as doing so would require 4.8 Earths.

Before we run with this number, however, it is important to recognize the potential bias in this approach.

Live like the Swiss

How many planets would we need if everyone lived the lifestyle of a typical Swiss citizen?

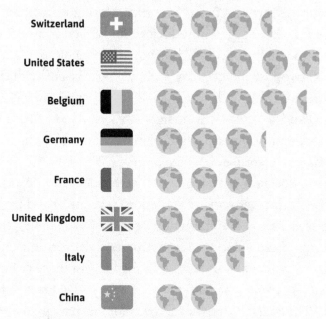

Source: *Le News*, 2016 based on the Global Footprint Network.

Bias in the Ecological Footprint Measure

Every measure has a bias, and Wackernagel was the first to call out biases in this ecological footprint approach. Advocates of the ecological footprint approach argue that it underestimates humanity's demands on the available resources of the planet. That is, they assert that it is a conservative measure of ecological impact. One implication of this assertion is that any estimate of population carrying capacity based on this ecological footprint methodology will result in numbers that are a bit high.

Criticism of this methodology comes in two forms. One set of critics focuses on the ecological footprint methodology's assumption that a given hectare of land can be used only for a single purpose, and that ignoring the potential of land to be used for multiple purposes biases the ecological footprint upward. In short, they believe humanity's ecological footprint is overestimated and that the case is not so dire.

The other group of critics, from the Breakthrough Institute, an independent non-profit research institute located only blocks away from Wackernagel's office, goes in the opposite direction. When they analyzed the Global Footprint Network's methodology, they took issue with how cropland and pasture are dealt with. In short, they object to the lack of a genuine measure of land overuse. The many ways in which agriculture degrades the land over time, limiting our planet's biocapacity, are well documented: Soil erosion. Overuse of water reserves. The accumulation of endocrine disruptors associated with fertilizers, herbicides, and pesticides. Instead, the Global Footprint Network's calculation of cropland and pasture measures only land area allocated to these purposes. Unfortunately, reliable and systematic data on these subjects simply do not exist for developing countries, as they do in the developed world. In the view of these critics, in the global ecological footprint calculation, cropland use equals simply cropland availability. The analysis does not account for mismanagement and degradation, and says nothing about the sustainability of agriculture.

These same critics take less issue with how the ecological footprint accounting calculations deal with forests and fisheries, which the Global Footprint Network believes to be in balance or even in surplus. They reserve their remaining criticism for the final category of global hectares used in this analysis—that allocated to the absorption of surplus carbon emissions. When cropland, pasture, forests, and fisheries are all in balance or even in surplus, the Global Footprint Network's analysis finds that the remaining 6 billion hectare overshoot in our use of global hectares is tied to the land that we would need to use if we were to absorb all of the surplus carbon that humanity creates. We know that most of our carbon emissions are actually accumulating in the atmosphere and being absorbed by our ever-acidifying oceans. Thus this calculation, the 6 billion hectares of overshoot, is entirely composed of human emissions of carbon that have been converted into a proxy measure of land area—the land area that would be needed to serve as a carbon sink if we are to avoid the worst impacts of climate change.

In short, although the Global Footprint Network's approach to calculating the human ecological footprint helps us grasp how the combination of all of our demands exceeds our planet's limited biocapacity, these critics have shown that the current iteration of this methodology lacks a certain sophistication in how it deals with important details, if only because of the lack of available global data sets. Nevertheless, it would seem that this method is a sound starting point for measuring ecological footprint geographically, even if it underestimates humanity's impact on our planet and provides only national granularity.

The Geographically Explicit Biocapacity Thesis

Although Wackernagel's method gets us to some geographical grounding, his accounting leaves us still searching in our quest to understand how humanity has affected the ecological carrying capacity of Earth at a scale granular enough to map it to each of our planet's various ecoregions. Equally important, Wackernagel's approach leaves us unable to geographically account for the ecological impacts of the increasingly urban existence of our species, at an urban level.

A sympathetic departure from Wackernagel's work is that of Professor Paul Sutton and his band of remote-sensing geographers, ecologists, and economic geographers. Their project, which I (not they) dub the "Geographically Explicit Biocapacity Thesis," sought to find a spatially explicit proxy variable for the human footprint that could be informed by more granular remote-sensing data.[67] Rather than national-scale estimates

---

[67] Sutton and his colleagues never formed a single name for their larger project, which was a decade-long evolution from insights driven by remote-sensing data to historic breakthroughs in how to understand the human footprint. In 2001, Sutton and his colleagues began implementing what they called a "Census from Heaven," estimating the global human population using nighttime satellite imagery. Building on this work, in 2002, they began building global estimates of market and non-market values derived from

of ecological footprint, Sutton and his colleagues sought to provide satellite-based estimates of constructed area on a 1 square kilometer (km2) scale as a spatially disaggregated proxy for the human ecological footprint. In so doing, they developed a global grid of constructed areas characterized by remotely sensed impervious surfaces. By building a set of algorithms that could differentiate between impervious (rooftop, sidewalk, parking lot, roadway, etc.) and not impervious (lawn, park, golf course, etc.) in remote-sensing satellite data, they developed a calibrated value of the percentage of impervious surface within each square-kilometer pixel. They then took high-resolution population data and divided the constructed area by population count, generating a disaggregated grid. This allowed them to estimate the constructed area per person, from a granular 1 km^2 unit all the way up to subnational administrative units, and ultimately to a national scale.

This method was not only a sympathetic advancement of Wackernagel's ecological footprint efforts. It was also a critical response to the equations driving Ehrlich and Holdren's energy thesis of Earth's carrying capacity. Ehrlich and Holdren's I = PAT equation, which appeals to many who recognize that population and consumption have an ecological impact, is nevertheless complicated by the difficulty of quantifying the role of technology. Sutton and his band cut to the chase and offered a simplified, spatially explicit, proxy

---

nighttime satellite imagery, land use data, and ecosystem service valuations. This led to the creation of the Empirical Environmental Sustainability Index in 2003, derived from the same body of data. By 2007, they were able to construct a Global Distribution and Density of Constructed Impervious Surfaces data set, which led to the 2009 article "Paving the Planet: Impervious Surface as a Proxy Measure of the Human Ecological Footprint." This stream of work culminated in the 2011 article "The Real Wealth of Nations: Mapping and Monetizing the Human Ecological Footprint." Although short on branding efforts, this group of scholars unlocked a powerful toolset for understanding the challenges associated with how humanity might better co-exist with the planet that sustains us.

measure human impact on the environment in their high-resolution geographical measure of "pavement."[68]

The Historical Truncation of Measuring Pavement

As we reach toward more granular geographical understandings of how our world has changed over time, we find ourselves with truncated source data. After all, Landsat, our first remote-sensing scientific satellite, only took orbit in 1974. Arguably, the state of the art of with respect to mapping the built environment is the Global Human Settlement Layer (GHSL) data set, developed by the European Space Agency and the European Union Joint Research Center. The GHSL data set has produced four spatially explicit snapshots of the extent of human settlement (1975, 1990, 2000, 2014) over the entire globe, using machine learning on a wide array of satellite imagery. Certainly a detailed analysis of global land records would enable the reconstruction of a historical global grid back into the earlier parts of the 20th century. This would be interesting and useful for understanding how core ecosystem goods and services from various ecoregions have been diminished over time. Yet for the purposes of bringing critically needed geographic detail to Wackernagel's ecological footprint endeavor, such snapshots through time prove very useful in helping us allocate the known national footprint accounts to these disaggregated 1 km² footprint delineations over time.

Biases in Measuring Impervious Surface

Whereas Wackernagel's effort is biased toward measuring the employment of ecological resources for agricultural production, Sutton's

---

[68] By "proxy" we mean that their work was specifically calibrated against Wackernagel's ecological footprint calculations, at the national and global scales. In so doing, Sutton and his collaborators inherited the biases of the ecological footprint methodology.

effort is biased toward measuring urban and suburban development. With a majority of humanity now living in urban areas and more than 75% of the world's population being projected to live as urban dwellers by 2050, this bias seems inherently useful when calculating ecological carrying capacity. Also, this approach certainly helps capture the subnational administrative districts featuring populated areas with modern infrastructure.

It will not, however, capture all of the human impacts, as a focus on impervious surface will miss all of the green groundcover that humans often replace native ecosystems with as part of their development of the landscape. Similarly, this bias will not capture the human-produced ecosystem impacts born of clearing, farming, and pastoralization. In theory, the proxy measure for urban impervious surface itself accounts for all of the agricultural land that is under cultivation in order to support these populations. However, clearly there are populations in lesser developed areas not characterized by impervious surfaces, but by subsistence farming or even transitional ecosystems. These populations and their ecological impact will be undercounted using the proxy measure.

Now, once we are operating at the geographically explicit scale set forth by Sutton et al., similar remote-sensing methods could also be used to determine other forms of human ecological impacts, with similar geographic granularity. Satellite remote sensing has demonstrated its effectiveness at identifying any manner of human-induced changes on Earth's surface at very granular levels. But this is a massive Big Data challenge that would require even more significant resources than those applied to generate the GHSL data set. If humans' transformation of the global landscape, regardless of its type, were measured geographically at a very granular level, Wackernagel's ecological footprint effort would have at its disposal the resources required to comprehend humanity's ecological impact at very local scales.

A Macroscopic Measurement of Carrying Capacity

Such geographically explicit proxies enable a sort of worldwide macroscope with which to monitor change and to engage in geographical accounting of ecological burden at a very fine scale. The Global Ecological Footprint analysis presented a clean, if geographically rough, accounting of the carrying capacity of the planet. It determined that, if we all lived at the level of the modern Swiss citizen, Earth could support only roughly 2.4 billion people without accruing further long-term ecological debt. Of course, this does not include the existing ecological debt that must be paid down.

Sutton and his team, with their focus on (sub)urban impervious surface, focused on humanity's impact on a much smaller geographical portion of Earth's surface. Yet as their analysis was intentionally calibrated as a proxy to Wackernagel's work, their beginning calculation for Earth's carrying capacity was the same—2.4 billion—well, assuming we were all modern industrialized Swiss citizens.

Yet these calculations gloss over humanity's larger footprint of ecological destruction as it relates to accumulated persistent wastes. Wackernagel's analysis really only includes humanity's vast well of carbon exhaust in its waste calculation. Indeed, the land-based global hectare methodology calculates that humanity currently demands 1.5 Earths mostly because of the amount of land that would be required to absorb the atmospheric carbon that we have produced over the past century or more. Wackernagel, and therefore Sutton et al., do not incorporate any of the other wastes outlined in Chapter 5. Once we are operating from a geographically explicit framework, we can take our geographically explicit understanding of this wide array of accumulated persistent wastes, superimpose it onto Sutton's global 1 km^2 grid, and understand how their calculations actually overestimate the carrying capacity of the planet.

The P3B Thesis

I view my own thesis as a sympathetic twist on the important work of Wackernagel, Sutton, and their kindred spirits. I firmly believe that Sutton's more explicit, fine-grained geographic approach is necessary to understand how humanity's ecological footprint affects specific ecoregions and the ecosystem goods and services that they generate. I also firmly believe that its focus on the impervious surfaces of an increasingly urbanizing world is a critically important lens with which to understand the ecological impacts of humanity's industrialization of Earth's surface.

However, as should be clear by this point in the conversation, I feel that both of the aforementioned methodologies systematically undercount the accumulated persistent wastes that humanity has created in an accelerated fashion for more than two centuries. As such, Wackernagel's 2.4 billion estimate of Earth's carrying capacity appears to be far too forgiving.

Another complicating factor is which "lifestyle" we baseline our estimates against. Wackernagel's 2.4 billion is based on a Swiss lifestyle, which is a very modern, affluent one. Yet the average Swiss person's ecological footprint is less than 70% of that of the average American. This means that countries that enjoy more affluence and consumption than the Swiss would need to reduce their consumption or their ecological footprint down to the Swiss level for his 2.4 billion number to be accurate, as the growing global middle class becomes more affluent.

To be fair to Wackernagel, the ecological footprint methodology is not intended to determine how many people Earth can support. It is a much more sophisticated and nuanced treatment of how nations can think about, and account for, their net contribution to the ecological burden that humanity places on our planet's biological capacity. It is I, not Wackernagel, who focused on the 2.4 billion estimate that is based on his calculation for the Swiss lifestyle. He simply used that estimate to communicate to his Swiss brethren the extent of their own ecological burden on our planet, pointing out that the Swiss impose a significantly larger per capita ecological burden than the average European (though not as bad as the Belgians!).[69]

Remember, Wackernagel's methodology, by his own admission, underestimates the ecological burden of humanity. Moreover, credible critiques of his approach indicate that this underestimation is considerably more substantial. So, for the sake of discussion, let us assume that Wackernagel's 2.4 billion is actually something more like 1.5 billion.[70]

I would still argue that this 1.5 billion estimate is wildly rosy, since it underestimates the accumulated persistent wastes that humanity has put into our various key ecoregions over the last century or so. The problem posed by these wastes is so acute at this point that an immediate, consistent decrease in population is needed in order to avert ecological catastrophe.

As a technological optimist, I would love to point out that humanity could very easily and very quickly increase our efficiency, decrease our wasteful by-products, and reduce our per person ecological footprint.

---

[69] This discussion draws from a 2016 article entitled "How Many Planets Would Be Needed If Everyone Lived Like the Swiss?" in *Le News*, which provides a lay-accessible discussion of the Global Footprint Network's methodology, as presented to Swiss readers. Since few in the developed world would consider it a burden to live as the Swiss do, that lifestyle baseline seemed like a constructive entry point into a quantitative discussion of our Earth's long-term ecological carrying capacity.

[70] To be clear, as an American, I am inclined to use the American standard, which is roughly 1.5 times as bad as the Swiss. Whereas humanity would require some 3.3 Earths if we were all Swiss, we would require some 4.8 Earths if we were all American. This approach would bring us to a maximum number of 1.65 billion humans living as modern Americans, as compared with Wackernagel's 2.4 billion living as modern Swiss, on the basis of today's global "production function." Thus, if we took into account the additional underestimation of humanity's ecological footprint, as discussed above, and discounted this 1.65 billion figure, we would end up with a calculation of Earth's carrying capacity that comes in at just about 1 billion, if we lived as modern Americans, who are the most wasteful population on Earth. And this is before we account for the wide array of accumulated persistent wastes that we now contend with due to humanity's profligacy. But let us continue to assume a global Swiss standard, for the sake of this argument.

However, before getting carried away with the latest and future prospective innovations, I would reiterate that the ecological weight of our accumulated persistent wastes is presently at a breaking point.

Under the current global production function, a population of 1.5 billion would still continue to generate persistent wastes that would accumulate over and above those legacy wastes, which will take generations to work off. And 6.5 billion of our existing 7.5 billion people will not immediately disappear in any ethical regime focused on our planet's ecological future. This means that even if the population steadily decreased from our present population of just over 7.5 billion (and more realistically, it will top out at least 9 billion), we will have wrought such ecological destruction and accumulated such suffocatingly persistent wastes that we would need to bring the population well below the Wackernagel estimate for generations, if we hope to dispatch these persistent wastes. Thus, at the current global production function, I suspect we would be fortunate to enjoy a population of 1 billion, living a modern Swiss lifestyle.

Any new technological innovations that might help us address our massive accumulation of waste would be a net improvement to the ecological sustainability of the current global production function. Luckily, both Wackernagel and Sutton are operating on the assumption that our average production function would look about the same in the future. If human history has taught us anything, it is that technological advancement is a constant.

Yet if we were to be technological optimists and assume that over the coming decades we could make humanity more efficient—producing less waste, becoming less ecologically impactful—and radically rewild much of the planet while extracting accumulated persistent wastes, one could imagine perhaps as many as 3 billion people inhabiting Earth—again, living a modern Swiss lifestyle—without accruing long-term ecological debts as we did in the 20th century and beyond.

Why not more than 3 billion? The finite nature of our planet's geography simply will not allow it if we actually seek to live within its long-term ecological carrying capacity. At least, not given the ecological debt that humanity has accrued over the past couple centuries, and the past half-century in particular. Over the long run, perhaps the number could be higher, but only after humanity fundamentally transforms its production function, and after much time and effort is spent working off the accumulated wastes and after rewilding much of our industrialization of the planet's surface.

Achieving a Geographic Accounting of the Planet

None of this will be possible if we do not first generate a shared biogeographical understanding of the ecosystem goods and services that our world's ecoregions are capable of generating and balance that understanding against the needs and wastes generated by modern industrial human civilization.

Our industrialization of the global landscape has not only been a story of humanity's destructive tendencies. It has also been the foundation of our modern human civilization that now stands at more than 7.5 billion souls. We have undertaken many forms of extraction and combustion, including deforestation, in order to fuel our many endeavors. We have cleared large swaths of our continents' historical habitats in order to cultivate bountiful crops and pastures that feed us. We have built manufacturing capabilities with insatiable appetites for natural resources in order to create goods that improve our quality of life and satisfy our acquisitive natures. We have built infrastructure to extend our reach across the farthest frontiers to ease the movement and communication of goods and people, though this has also made it as easy as possible to exploit and denude each new frontier. And we have built cities that have increased the quality of human life immeasurably, at the cost of laying waste to whatever historical ecosystem once lay within each city's limits.

This is to say that we, as a species, require each of these forms of industrialization that we have wrought on the global landscape in order to live our modern lives. I, for one, value many dimensions of modern life and do actually judge many of them as useful progress, though I (obviously) have become very concerned about modern life's ecological consequences. The more humans there are on the planet, the more of the planet must be industrialized to one degree or another, even if sustainability strategies were taken to their logical extremes. However, each of these activities has also profoundly undermined Earth's ability to support those of us who inhabit this engineered landscape. These activities have undermined Earth's carrying capacity by annihilating, crippling, or hobbling historical habitats that used to generate vital ecosystem goods and services for humans and the food chains we occupied.

Our industrialization of the global landscape has also undermined Earth's carrying capacity by generating wastes that Earth is left to carry, while being expected to generate the same ecosystem goods and services that it always has. Indeed, as the population has grown, we have, in many cases, expected our ecosystems to generate even *more* goods and services, even as we spoil these ecosystems at an increasingly alarming rate. The portions of Earth spoiled by human wastes are concrete, measurable geographies—as has been discussed in the previous chapter and quantified in rigorous digital geographic data sets.

The question is how we might achieve a balance that allows our species to continue to thrive, as modern humans inextricably tied to some form of industrialization, while enabling the long-term ecological viability of our planet.

The achievement of such a balance will require a rigorous geographic accounting. The extent of the global landscape's industrialization and the ways in which humanity has laid waste to the planet can both be accounted for geographically. The world's ecoregions can also be accounted for geographically. And then the former can be subtracted, geographically, from the latter. As one juxtaposes the geographic growth of the former (industrialization and waste over the global landscape) with the spatio-temporal distribution of the past century of population growth, one can see that modernity has demonstrated a certain per capita geographical footprint in each ecoregion.

Each additional person demands more roads for transport and more impervious surfaces on which to park their vehicles. They demand more buildings in which to work, learn, shop, practice their faith, and perform all the other functions that modern humans require. Each additional human also demands more croplands and pastures to feed him or her. Each additional human demands some amount of space for his or her wastes to accumulate, or to be recycled either by humans, by machines, or by nature, over some period of time. Yet each additional human will still demand that Earth naturally generate marginally more ecosystem goods and services to support them with sufficient clean air, fresh water, and the like.

At some point, the demands of additional humans collide with each other. At some point, there are no longer enough undisturbed ecosystems to generate the range and volume of ecosystem goods or services needed to sustain basic planetary processes. At least, there are not enough while these ecosystems are also being actively harvested to continuously feed the demands of human industrialization, and while these ecosystems are being weighed down by humanity's accumulating wastes. This collision of demands results in ecological debt. This ecological debt can be understood only when accounted for geographically.

Putting Our Ecological Debt in Modern Historical Context

This debt began to clearly manifest by the middle of the 20th century, as the world population reached 3 billion. The obvious manifestations were in the developed world, where the industrialization of the global landscape had over a century-long head start, and total war had wreaked ecological havoc across broad swaths of the planet.

However, from then on, even as the developed world rebounded ecologically in important ways, the growing global population, in both the developed and less developed parts of the world, accumulated more ecological debts. From 1950 until now, older forms of human industrialization of the landscape—forms rooted in the 19th century, and even before—expanded considerably. Many colonial plantations predated this period, and these and other historical development methods expanded massively in the mid-19th century to satisfy global demand. Dated but effective agricultural and manufacturing practices stretched all across the planet to ecoregions that had previously gone largely untouched, both fueled by and fueling the global baby boom of the mid-20th century. This industrialized large parts of historical ecosystems while also generating increased waste loads that are now accumulating in previously pristine ecoregions.

In addition, new forms of industrialization and waste production that were only nascent in the late 19th century rose to prominence by 1950, and then kicked into high gear. Gasoline and combustible engines enabled transportation networks to be forged to distant frontiers by any industrious individuals willing to carve a path through an ecosystem. They also enabled unimaginable growth in global cargo ship transport and global air travel. And they made it profitable to plow, sow, and reap crops on an epic scale to not only generate food, but also harness agricultural crops as material inputs to industrial chemical engineering processes. The production of nitrogen fertilizers, plastics, radioactive materials, and new forms of endocrine disruptors also changed the game, as they were broadly applied to both the developed world and the distant reaches of the developing world. Not only were there more manufactured materials spread across an increasingly industrialized landscape of increasing geographic extent, these materials were also orders of magnitude more toxic and more persistent than previous forms of waste, leading to shocking levels of toxicity accumulating in the environment.

In short, the global population of 3 billion that existed just after 1950 sat atop an enormous ecological debt that had already been accrued, primarily by developed nations—nations that had also ecologically compromised the planet through total war. Yet rather than holding the line on population and working aggressively to pay down this ecological debt, the human species actually increased its overall fertility rate, while simultaneously unleashing new, more ecologically destructive forms of industrialization of our planet's global landscape, now extending to parts of the planet that had previously gone largely unaffected ecologically by the industriousness of modern human civilization.

To be clear, the industrialization of the global landscape was not the goal of modern humans. Industrialization was a means to an end. Much of this was in the service of progress—the same forms of progress that the developed and developing worlds alike continue to pursue. Together, they improved humans' quality of life. Industrialization enabled many forms of progress, including decreased maternal mortality, decreased infant mortality, increased longevity, reduced incidence of disease, protection from natural disasters, and relief from drought and famine. And it brought us all manner of modern miracles, from air conditioning and refrigeration to public health interventions and modern surgery, to telecommunication and air travel to computers, the Internet, and the World Wide Web. Humanity has engineered vast geographies where such things are possible. And many in the developing world yearn for these same forms of progress, even if it is at the expense of vital ecosystem goods and services.

But this massive cultural project called "progress" has not accounted for the role of wildernesses in humanity's existence on Earth. For so long, wilderness was something to be tamed, beaten back, or harnessed. Demolishing wildernesses in the service of civilization was key to modernity, until the conservation awakening of the American Progressive

Era. Though the terms "sustainability" and "resilience" have modified our concepts of progress to be less ecologically destructive, they will never reconstruct the idea of "progress" into an ecologically friendly concept until we appreciate that wildernesses are essential to humanity, and that Earth can support only a limited number of modern, industrialized humans—even if we are energy efficient and ecologically conscious.

The techno-optimist in all of us would argue that human ingenuity will invent new forms of landscape industrialization and new forms of waste that are less impactful on the planet, making it possible for Earth to carry a larger population. It seems quite reasonable that such ingenuity could have helped pay down the ecological debt accrued by 1950. It would have been possible to harness this ingenuity to somehow transform the ecological footprint of 3 billion humans to fall into balance with the ecosystem goods and services of what remained of our ecoregions, had we nurtured them back to health. However, the trifecta of forces that characterized the second half of the 20th century—namely, new forms of industrialization of the global landscape, the generation of new persistent forms of waste, and the geographical expansion of development to the most remote remaining ecoregions—has generated an unspeakable ecological debt that is far beyond what human ingenuity might be able to overcome, given the ecological footprint of the more than 7.5 billion people living with the benefits of the modernity that they seek.

The Geographic Trade-Off

In actuality, the issue is one of physical space: geography. Even if we could create a modern human way of life that allows elbow room for the next impending waves of human innovation, that is more energy efficient, that generates fewer waste by-products, and that relies on fewer hectares of industrialized planet per human, there would need to be vastly fewer humans on our planet in order to restore some sort of meaningful ecological balance that could sustain humanity over the long term.

Rather than confronting this hard reality, many still argue that everything will be fine, ecologically, as the world approaches a population of 9 billion or 11 billion. Some argue that densification through urbanization can absorb the increased population. Some argue that there is plenty of land on a large planet to accommodate the increased population, using existing land use patterns. None account for the waste that these populations will generate, the additional industrialization of the landscape they will demand, or the amount of the planet that would need to be set aside in order to ensure that the volume and variety of ecosystem goods and services needed in order to support such a large population are in fact generated.

They do not account for these issues geographically. They do not show their math. They just generate a cacophony of conflicting rationales that only agree on the notion that no further thought is needed and no action is required. It is a curious form of myopia that they advocate. Asleep at the switch, they are prepared for the consequences of inaction to crash down upon future generations.

If you take issue with the new, lower population plateau of 3 billion humans that I suggest, and you seek to create your own estimate, I would challenge you to think about the real geographic trade-off that humanity foists on our planet's ecological carrying capacity, which in turn supports our species. In your calculation, what population could be supported (a) without undermining our atmosphere, climate, and oceans with excess human-produced carbon; (b) without generating excess nutrients that create ocean dead zones; (c) without accumulating more persistent toxic pollutants; (d) without disturbing biophonic processes with excess anthrophony; and (e) without having garbage inundate our high seas? And how far would global population need to be reduced in order to allow these accumulated persistent toxic wastes enough time to break down or to be removed? Also, what population could be supported (a) without extracting and combusting more materials that spoil historical ecosystems and alter our climate, (b) without having agriculture suppressing larger tracts of former ecosystems, (c) without demanding more natural resources for manufacturing than Earth can replace in comparable time frames, (d) without infrastructure that transects ecosystems and generates impervious surfaces where once lay productive ecosystems, and (e) without allowing urbanization, suburbanization, exurbanization, and informal settlements to annihilate or burden additional key ecosystems?

Due to the persistence of pollutants, human waste generation would need to drop to zero and remain at that level for several generations in order for Earth to rebound from the current toxic burden. Due to the way in which agricultural production has compromised ecosystems, the geographic area allotted to it (in terms of square kilometers) would need to drop precipitously to allow their restoration. Due to the ecological devastation associated with the informal settlements of the developing world, such population centers would need to be depopulated by people migrating to sustainably engineered urban centers. Due to the ecological productivity of many land-water boundary areas, significant populations and portions of the industrialized landscape would need to move away from coastal, riverine, and inland bodies of water. Due to the insatiable demand for raw materials and their inevitable chemical by-products, manufacturing activities would need to decline precipitously in order to reduce the draw on natural resources and the production of persistent pollutants, and the resulting postconsumption waste.

Not only do I challenge you to produce your own estimates and to show your data geographically over time, I welcome it. Only through robust debate will we be able to navigate our way to a new, sustainable future. That the world population will drift upwards to 9 billion is a near inevitability. At 9 billion, the most aggressive embrace of the Sustainable Development Goals (SDGs) recently passed by the United Nations and the global community would still not get us close to a net-zero waste profile and would not reduce land usage. Indeed, land usage and its despoiling will likely increase. Our oceans will see further plundering and destruction. Our atmosphere will continue to be poisoned. The ecological debt of humanity will simply get larger. And even if humanity succeeds in coaxing itself into population decrease, it will be some time before it can bring its waste profile to zero, to begin letting the accumulated persistent wastes recede. Even when the waste profile is brought to zero, reaching the point at which ecological restoration and rewilding will return enough of the planet to meet E. O. Wilson's 50% goal will require an even smaller population—again, even with the most aggressive embrace of the SDGs.

CHAPTER NINE

How to Bring the Population Down to 3 Billion

―――――――

This is not just another book on overpopulation. It is about the carrying capacity of Planet Earth and how humanity can bring our total population in line with our planet's ability to support us. As such, it requires all of us to think about our world, and the challenges that it faces, a little bit differently. The planet that has nurtured our species over the past 200,000 years has been ravaged by humanity's industrialization of the global landscape over just the past couple centuries, and it is struggling to support us. It could support a human population of 3 billion. The stark question in front of us is how to bring our more than 7.5 billion (and fast approaching 9 billion) population down to this level. Many actions and innovations will be required just to make the 3 billion number a legitimately viable option, and indeed to actually bring the population down to that number. First, though, we must reverse runaway population growth.

In short, it is a question of women's empowerment. To be clear, women's empowerment is an inherent good, in and of itself. It should not be viewed as a means to an end. Women's empowerment, education, and autonomy over their bodies are things we should support for their own sake. Yet they are clearly central to how our species can align with the long-term sustainability of the planet that supports us. If women across the planet were educated and integrated into the workforce, and yes, given access to meaningful family planning technologies, total fertility would plummet. Economists such as Jeffrey Sachs, in his book *The End of Poverty: Economic Possibilities for Our Time*, have documented this phenomenon in specific nations. And economists are not alone in this observation.[71]

The 19th and 20th centuries are a lesson in half-blindness. While we aggressively used technologies to achieve many forms of progress that resulted in lower infant mortality, lower maternal mortality, better public health, and greater overall longevity, we completely failed to empower women around the world equally and give them access to family planning technologies to help keep the global population in check. While we invested enormous resources between the 1930s and the 1970s in the so-called "Green Revolution" that increased agricultural production worldwide, enabling reductions in death through privation and resulting in population growth, we did little to match this effort with investments in women that would allow them to be managers of their own fertility. The result was the runaway population explosion that we have experienced, paired with rapidly rising living standards, first in the developed world and then spreading quickly to the growing global middle class. This population explosion, together with the growth in per capita ecological footprint that modernity has unleashed, has brought a wave of ecological destruction only because so many women around the

---

[71] As this discussion progresses, you will see that women's empowerment cannot be seen in isolation from other contextual variables such as poverty, levels of urbanization, socioeconomic options, cultural norms, and the like. Women's empowerment in rural Sub-Saharan Africa may play out quite differently than in heavily urbanized parts of Latin America. As such, women's empowerment cannot be seen as an overly broad-brush panacea. Yet, with the rapid rise of the global middle class in so many historically less developed nations, the time is coming fast when the vast majority of women will live in modern, urban, industrial contexts. And, despite factors that might militate against women's empowerment in traditional rural villages and other contexts, examples abound of how women's empowerment strategies can alter fertility patterns in these contexts as well.

globe were never given opportunities for education, empowerment, and self-sufficiency, particularly in the realm of reproductive freedom. As a result, an ever increasing number of men, doing what men had done for the prior 200,000 years, planted seeds whose fertilization led to global ecological destruction.

In order to navigate our way to a new, lower population plateau, we must develop a shared understanding of how global fertility works and what actions individuals can take to bring our population into balance with our planet's ecological resources. Let us first discuss how fertility is calculated, what it takes to achieve replacement fertility levels, and how we might achieve below replacement level fertility in a way that can put humanity on a glide path to a new, lower population plateau of 3 billion.

Calculating Fertility

It has long been said that you cannot manage what you cannot measure. Although fertility is measured and forecasted by all manner of public agencies and academics, the measures they use are cloaked in jargon and advanced statistical methods that make them all but impenetrable to a lay audience. The result is that those individuals who might seek to undertake family planning to help achieve a globally sustainable level of fertility have no practical guidance.

The term "total fertility rate" (TFR) is the high-level statistic that most focus on when looking at how nations are doing with regard to runaway population growth, or in the achievement of replacement fertility levels, or even negative fertility rates.[72] The TFR is defined as the average number of children that would be born to a woman over her

---

[72] For these purposes, think of "replacement-level fertility" as the TFR at which a population replaces itself from one generation to the next, without accounting for migration. It generally refers to a TFR of 2.1, or 2.1 children per woman.

lifetime if (a) she were to experience the current age-specific fertility rates throughout her lifetime, and (b) she were to survive from birth through the end of her reproductive life. This means that the TFR encompasses issues such as a country's infant and child mortality rate and its maternal mortality rate. It is an amalgam of many factors that clearly glosses over much social complexity.

As such, it is hard to tell women in a specific country that if, on average, they each had 2.1 children, over the course of their lifetimes a particular population outcome would occur. Particularly in nations where infant mortality or maternal mortality rates are high, and concerted action is being taken to bring these numbers down, the resulting population outcome could be substantially higher numbers of people. Nevertheless, tracking a country's TFR is currently the best way to monitor its progress toward sustainable population practices.

Understanding the factors underpinning these population dynamics is another question entirely. How these factors manifest over various geographies and conspire to shape the TFR of a given country or locality is a complicated process, thus spawning the field of spatial demography. It is the overall goal structure that a given society and its constituent communities and citizens might seek that will shape its TFR. Runaway population growth, steady population growth, replacement-level fertility, or negative fertility rates can be the result, depending on the choices humanity makes.

Achieving Replacement-Level Fertility
History has shown that societies that hover around replacement-level fertility rates rather reliably boost their economic development, alleviate poverty, increase social equality, and reduce agricultural pressure on their ecosystems, their water supply, and their local climate dynamics. Many might assume that if such good things befall societies that reach for replacement-level fertility, there must be some hitch. The default assumption is often that such an achievement must require difficult and

painful strategies. Actually, the most effective approaches to achieving replacement-level fertility are not coercive. These approaches advance gender equity, promote economic growth, and can save millions of lives.

In general, where children are considered an important part of the labor force, as in developing nations with substantial rural areas, fertility rates tend to be higher. Conversely, people living in urban areas, whether in developed or developing nations, tend to have a lower TFR due to the lower demand for child labor, the higher cost of raising and educating children, and the broader access to family planning technologies. Where women have educational and employment opportunities outside of the home, TFRs tend to be low. Indeed, in developing nations, women with no education generally have two more children, on average, than women with a secondary school education. In geographies where infant mortality rates are lower, women tend to have fewer children. Where women get married later, the TFR is lower, and where women's average age at marriage is 25 or older, their TFR as a cohort is lower. Where there are private and public pension systems or reliable retirement regimes, TFRs tend to be lower, as parents see less of a need for an abundance of children to care for them in their old age. Where there is access to reproductive health services, legal abortion, and reliable family planning technologies—in other words, where women are able to achieve control of their reproductive destinies—the TFR is markedly lower.[73] Obviously, geographies over which many of these factors overlap enjoy replacement-level or even negative fertility rates.

Importantly, religious beliefs, traditions, and cultural norms paired with opposition to family planning technologies (an opposition which is often induced by these religious beliefs, traditions, and cultural norms in

---

[73] In 2016, the UN Population Fund, in a report entitled "Universal Access to Reproductive Health: Progress and Challenges," calculated that 350 million women in the poorest countries did not want to have their last child, but did not have the means to prevent the pregnancy.

the first place), can conspire to counter all of these factors and lead women to favor or at least tolerate large families.[74]

Embracing Below Replacement Level Fertility to Achieve a New Lower Population Plateau

So often, alarmists bring to our attention those nations that are facing the dangers of stagnant or negative fertility. These alarmists are often, if not exclusively, economists. Neoclassical economists, trained in the orthodoxy of thinking that economic growth is tied fundamentally to population growth, cannot imagine their way to a world of declining populations capable of navigating themselves toward economic prosperity. These alarmists talk of the need for such nations to import immigrant labor to undertake key labor categories or to take care of an aging population. Japan and European nations, notably, are the targets of such analyses. Some even advocate for state-based compensation schemes to drive couples (or even single women) to procreate, so called "pro-natalist" policies. This, in the face of the near vertical population growth on our planet.

However, if one traded this narrow neoclassical economics mindset for an ecologically minded worldview, it would become clear that Japan, Europe, and even the United States (which would enjoy negative fertility if not for immigration) should actually be embraced as models of how to achieve a new lower population plateau. The population discussion of the past two decades has been deeply schizophrenic, at one moment denying the possibility of bringing population growth under control and at the next moment decrying the impending societal collapse that a given nation's negative fertility might induce. This is simply ill-informed silliness.

---

[74] For a more detailed discussion of how societal factors interact to shape TFR, and which combinations have led to the real-world achievement of replacement-level fertility, one could refer to a series of reports by the World Resources Institute entitled "Achieving Replacement-Level Fertility."

The truth is that educating and empowering women, in the modern industrial context, leads inevitably to a decline in fertility and even negative fertility. If we take seriously the ecological predicament that humanity's population explosion and industrial expansion of the past century or two has wrought, we should shake off our current schizophrenia in favor of a reasoned campaign for global women's empowerment, women's education, women's workforce integration, and women's access to family planning technologies. It is actually that simple. Yes, there are cultural barriers to be overcome, but I would argue that dispelling the economic imperative born of the neoclassical economics mindset is the biggest barrier. Alternative economic concepts can lead to economic prosperity and continuous economic innovation without embracing the imperatives of an economy predicated on an unsustainable ecological-resource-consuming juggernaut caused by runaway population growth.

Why Some Fertility Strategies Succeed, and Why Some Fail

Study after study has shown that when women receive education (both about their family planning options and in general) and when they have access to various family planning technologies, they choose to have fewer children and to start having them later in life. Some would argue that religion and other cultural factors are insurmountable barriers that will always prevent such behaviors. However, there is evidence to the contrary.

Costa Rica is an example of a Catholic country that dramatically lowered its birth rate in a short amount of time through rather simple measures. In 1962, the Evangelical Costa Rican Alliance organized "Good Will Caravans" that drove to rural communities and offered birth control devices and vasectomies, among other health services. This effort began to spread, and in 1968 the conservative Catholic president of Costa Rica, José Joaquín Trejos Fernández, ordered all government-funded health facilities to offer contraception. The TFR in 1960 was 7.3 and by 1975 it was 3.7. Today it is 1.8. This steep drop was achieved through education about and access to contraceptives.

Costa Rica is not a unique example. Other countries such as Iran and most countries in Europe have similar success stories. The fact is this: when women receive education and when they have access to various family planning technologies, they choose to have fewer children, starting later in life.

Some nations' fertility strategies have clearly failed. But the reasons for these failures were clear. Too often, they did not encourage children to not have children. They simply imposed a per person reproduction limit, such as China's notorious "One Child" policy. This led to culturally induced, gender-based abortions, which then led to a disproportionate number of heterosexual males with insufficient numbers of female mates.

Instead, policies should be firmly based on the empowerment of women and the encouragement of women to have fewer children, beginning later in life. The timing of birth, a fundamental factor that has been overlooked by population scientists, will have the largest effect. The dynamic operates much like the concept of compound interest. Population will compound faster, the earlier women have children. If women have control over the timing of family planning, this compounding effect will be curbed quite quickly. The side effect will also be a much more rapid accumulation of wealth and increase in quality of life within the societies that achieve such a purposeful throttle on their population growth at the individual household level.

Specifically, the solution is the creation of a new, global, species-wide norm that women not only have fewer children, but that they begin conceiving them later. This would replace the older norm that women begin conceiving children upon the onset of puberty. Children having children has been a source of population explosion simply due to the continuous compounding of population growth as infant and maternal mortality rates trend toward zero. Of course, the calculation of total fertility is quite complicated and has many dimensions. Yet in the end, there is a set of behaviors and norms that have proven that negative fertility rates can be achieved. Paradoxically, in many developed nations this trend toward negative fertility rates has many policymakers, theorists, and citizens at their wits' end. This is due to the conceptual prison that modern neoclassical economics has put us all in. This is the subject of our next chapter.

CHAPTER TEN

Reimagining Economics for an Era of De-growth

When it comes to the issues of population growth, the conceptual framework of neoclassical economics—the economics everyone has been taught in school for the past century—has put us all in a conceptual prison from which it will be hard to escape. It has led many a policymaker and businessperson to think that if an economy is not growing, in terms of gross domestic product (GDP), it must be dying. Perhaps, instead, we have hit a point where precisely because our economies continue to grow in this manner, our planet and our species are at risk of dying. It is high time that our notions of national economic growth be replaced with a focus on individual economic prosperity and wellbeing.

Anyone who has taken Economics 101 knows that one of the major components in all growth models is population growth. It is not that positive GDP growth could not occur if population decreased. After all, another major component of economic growth is productivity growth, borne of technological innovation. Still, if we entered a world where the number of consumers declined every year, the neoclassical economists would have us obsessively focus on "economic stagnation" or "economic contraction." For good reason, such "recessions" and "depressions" are the ultimate boogeymen for which economic prognosticators remain forever vigilant. The very definition of a recession or a depression is tied to a quarter over quarter decline in GDP.

The conceptual prison that is neoclassical economics has its roots in the same Industrial Revolution that unleashed the explosion in population growth that has wrought ecological havoc on our planet. As such, it should be no surprise that any journey to a new, lower population plateau will require us to embrace a new set of economic concepts that are less centered on overall economic growth and more focused on productivity growth, increased per capita income and quality of life, and the private and public finance mechanisms that will be required to enable smooth and resilient decreases in consumption and land use.

This last point will require a profound reimagining of economics since it will manifest in so many ways—not the least of which will be the ways in which growing real estate vacancies and landscape retrenchment could lead to the massive destruction of economic value, if not managed correctly. Managed poorly, or with an eye toward growth, this shift could have a crippling effect on capital markets, since so many stocks, bonds, and other financial instruments are premised upon some minimum amount of economic growth due to population growth. As we reach peak population (assuming we hit it before total ecological destruction is achieved), the basic premise of "economic growth" that proceeds apace with population growth will become curiouser and curiouser.

Interestingly, economists and economic observers are already grappling with related issues such as the robotification of society, its implications for the future of work, and the possible practical need for mitigations such as a universal basic income or more traditional social safety nets. The same dynamics that are driving these discussions would only be exacerbated by a systematic decrease in world population. Under a scheme of population decrease, aging populations would have far fewer of the younger generations to care for them and to support their financial social welfare—far fewer than the already alarming forecasts suggest. And this would go on for generations, until the new plateau is reached. Under a scheme of population decrease, the continued automation of work would still affect the working-age population disproportionately, as certain categories of work are eliminated or vastly reduced. Thus, while the number of people that might require a scheme such as universal basic income might decrease, such schemes would still remain relevant—unless all technologically displaced workers became providers of eldercare.

Re-conceptualizing economics in a world of population decrease will be an enormous challenge, which should begin today.

The Spirit of Ecological Economics

"Where conventional economics espouses growth forever, ecological economics envisions a steady-state economy at optimal scale," wrote Professors Herman E. Daly and Joshua Farley in their foundational economic textbook, *Ecological Economics: Principles and Applications*, published in 2004. This book introduced critical concepts to the field of economics, particularly the concept of "uneconomic growth."

Daly and Farley raised a central question about the massive blind spot in neoclassical economic thinking. Why is it that neoclassical economists have come to ignore the existence of marginal disutility, as they have embraced the concept of marginal utility? They always are cognizant of the potential marginal utility of growth but are blind to the increasing marginal cost of growth. Indeed, as natural capital is transformed into human-made capital, history has made it clear that sacrificed natural capital services and the disutility of labor represent real costs of growth in many scenarios. Fundamentally, neoclassical economists have ignored the marginal disutility (the increasing marginal cost) of growth, since their vision is of an infinitely large world filled with infinite resources. In neoclassical economics, there is no real notion of planetary limits. Their approach has been called the "empty-world vision." They might abstractly recognize the concept of uneconomic growth, but they claim that we are not yet at that point, and that because the marginal utility is still very large and the marginal disutility is still negligible, their focus should be on the former.

The more extreme version of this neoclassical worldview, which seems to drive a total neglect of the costs of growth, has economists wrongly viewing ecosystems as a subsystem of the economy, rather than the reverse. Failing to understand that the economy is a subsystem of the larger global ecosystem is an error of hubris that has led neoclassical economists to completely fail to determine value within the finite ecological context of the planet. Their failure to grasp the economy's "optimal scale" as based on a calculation of its ecological carrying capacity may simply be a reflection of the balkanization of the academy. Economists and ecologists, for too long, have collaborated far too little on issues of core theory. It may also be a reflection of economics' conceptual origination in the late 18th century, before the concept of "nature" was invented by Alexander Von Humboldt in the early 19th century, and the concept of "ecosystems" was invented by Sir Arthur George Tansley in the early 20th century.

If economists had nature and ecology as core concepts in their canon, it seems clear that they would embrace the notion of an optimal scale for the macroeconomy where marginal utility equals marginal disutility—the way Daly and Farley help us understand. Instead, it is common for economists to account for waste as just another resource that we have not yet learned to use. While over the long run this may very well be true, as John Maynard Keynes once quipped, in a very different context, "In the long run, we are all dead." Neoclassical economists would have us not account for the massive accumulation of persistent wastes as part of the marginal disutility of economic growth. Even as our finite planet drowns in these wastes, they would have us focus on the value added by human labor and capital, to the exclusion of natural processes.

Lastly, it is worth recognizing that the field of ecological economics does have its own blind spot in its intellectual commitment to a steady-state economy. In a way, ecological economists echo John Stuart Mill's mid-19th-century idealist concept of a "stationary state" (itself a reaction to the implications of Malthus's thinking), in which birth rates and death rates balance out, and production rates equal depreciation rates, so that the stock of people (e.g., population) and the physical capital stock would be in a state of dynamic equilibrium. Through a modern ecological lens, Daly and Farley would have us seek an economic equilibrium that enables the maintenance of our planet's ecological life-support systems far from the edge of collapse, which in their view require an end to material growth of the economy and a healthy, thriving human population that is focused on this collective goal of long-term ecological sustainability.

In the end, their theoretical commitment to stasis fails to appreciate the need for several generations of dynamic de-growth before our population and our economy can work in ecological balance with our planet. Perhaps this is because they believe that the world population has not yet exceeded the ecological carrying capacity of the planet, and that we could simply change our economic tenets and capitalistic behaviors and take a glide path toward such a dynamic equilibrium in the near term. On this, as is clear from the central thesis of this book, I could not disagree more.

There is no dynamic economic equilibrium that we can rely on to get to the lower population plateau required for the long-term sustainability of our planet and our species. Indeed, I would suggest that there is no such thing as economic equilibrium, other than a false economic assumption that neoclassical economists have unwittingly adhered to for generations, founded on the choice of maximization or optimization mathematics driving their theoretical models, rather than on empirical observations of how economies actually work. In the end, the insights of ecological economists need to find their home in a body of economic thought that appreciates the fundamental dynamism of long-term economic change.

The Economics of Technological Change

When the definitive history of 20th-century economics is finally written, it will be laid bare that the mathematical rigor of neoclassical economics, while powerful, elided many important concepts at the core of how economies actually work. Indeed, far more was omitted (or systematically de-emphasized) than the concepts of marginal disutility and optimal scale. This actually became clear to many economists by the last quarter of the 20th century. In the quest to understand what drives long-term economic growth, it was recognized by the likes of Robert Solow, for instance (see the Solow Growth Model), that technology plays a central role in driving economic growth, as it became clear that population growth, the accumulation of capital, and other key variables did not explain observed levels of economic growth.

This led several scholars to pioneer a field of economics sometimes called "the economics of technological change," which sought to understand the core economic processes at the heart of such change that so clearly drove long-term economic growth. These scholars quickly realized that many of the core assumptions of neoclassical economics would need to be relaxed or even replaced by more empirically rigorous theoretical models. They embraced many of the ideas of behavioral economists, long before mainstream academia came around to their truths. They also demanded a better appreciation of how information, knowledge, and learning actually worked in economic growth, decades before neoclassical economists came around. A third area where these scholars stepped out of the mainstream, and with which neoclassical economists have yet to catch up, is the rejection of notions of economic equilibrium.

Building on the concepts of Joseph Schumpeter, these economists recognized that long-term economic growth is not characterized by dynamic equilibrium. Instead, it is characterized by a process of "creative destruction" through which innovators continuously introduce new technologies, organizational strategies, and business models that create enormous amounts of economic value while diminishing, obsolescing, or destroying the economic value of previously dominant economic players. Such "neo-Schumpeterians" saw this process for what it was—a process characterized by the evolutionary dynamics of variation, selection, and path dependence in which investment in search (or research) activities yielded new innovations about which the market (whether consumers or business supply chains) could make very disruptive selections. Consumers could abandon horses and buggies for automobiles, but only after technologists and innovators invested in these new creations and built new organizations with new business strategies to carry these innovations forward.

In their seminal 1982 book *An Evolutionary Theory of Economic Change*, Richard R. Nelson and Sidney G. Winter brought together these observations in a new body of theory which offered a set of conceptual building blocks that would properly equip those who truly sought to understand long-term economic change—and the technological change at its very core. Nelson's larger oeuvre, which some dub "evolutionary economics," helped us understand how an economy composed of many interacting industrial sectors that change discontinuously could still demonstrate stability over time. This is due to the creative destruction caused by the co-evolution of each industry's core technologies, industrial structure, and supporting institutions. This intellectual project sought to understand what set of public policies and private strategies could help nations harness the power of technological change to achieve long-run economic growth. For, as is empirically clear to everyone, economies do not just magically grow at a maximal rate. This is why nations have long fiddled with the levers of fiscal policy, monetary policy, trade policy, labor policy, and, yes, innovation policy.

Economies are collective human endeavors that require innovators, business professionals, financiers, consumers, and policymakers to collaborate in order to achieve social welfare improvements as individuals and corporations seek to create their own profits and wealth. These economies can find themselves both buoyed and limited by forms of historical lock-in to particular paths of economic change. So-called "path dependence" allows economic incumbents to reap the returns of their investments. However, they must always beware of the new innovators whose creativity might just destroy the value captured by the incumbents.

All this is to say that economies are not characterized by equilibrium. If anything this is just an illusion sold to consumers, investors, competitors, and policymakers by economic actors who want to lull them into complacency. Thus, grounding the insights of the ecological economists in a conceptual framework based on equilibrium is ultimately a losing proposition. To get our global society "rightsized" at 3 billion humans will require the introduction of many new (hopefully more ecologically friendly) technological innovations that will, ideally, creatively destroy the more ecologically harmful industries while creating tons of economic value. And once we find our way to this new, lower population plateau, the ensuing continuous technological and economic discontinuities will have major natural resource implications.

It is clear that there is no such thing as a "steady-state economy." This does not mean, however, that the concepts of "optimal scale" and "uneconomic growth" (or the marginal disutility of growth) have no meaning. Indeed, they take on even more meaning in the context of evolutionary theories of economic change that properly grasp the economics of technological change as a central driver in long-term economic growth—or de-growth, as it were.

The Robots Are Coming

Let there be no doubt that a central trope in any scenario of economic growth or de-growth will be robotification of the future economy. If you are tuned in to the new and impending forms of automation, you are sure to understand that robots of many kinds will be a part of the fabric of our everyday lives in the coming years. As technological changes creatively destroy our old economy and serve as the foundation of our new economy, the question is whether these innovations will bring us a more ecologically sound future, or whether they will merely supercharge all of the bad habits honed by humanity during the 19th and 20th centuries.

Robots are defined as a constellation of sensors (that sense the world), actuators (that take action within the world), and processes (that seek to understand, react to, and coordinate between these two), working in concert for a single purpose. Most of our mental pictures of robots are of machines that resemble humanoids, like Robby the Robot from the 1956 movie *Forbidden Planet*. However, the robotification of the global landscape will happen in a multitude of very different ways over the coming decades, with disaggregated but networked sensors, actuators, and processes interacting in endless combinations, along with many types of (semi-)autonomous electromechanical objects. These robots will automate a blinding array of activities that are currently considered human endeavors.

What is fascinating is how much concern is voiced over the coming robotification of the economy at the same time that people voice concerns over the insufficient number of people available to navigate our societies through the rocky shoals of the future, due to aging populations and declining fertility rates in some nations. In the end, much of the consternation associated with the latter point is more of a financial concern, wondering who will pay to support the older generation in their golden years, if there is an undersized population of younger taxpayers. But why not frame this challenge as an opportunity for automation itself, a target for creative destruction? Innovators could introduce new technologies, organizational strategies, and business models that could help us navigate our way to a lower population plateau while providing our aging generations with superior care to that provided to previous generations of the elderly.

There are many other challenges of equal or greater magnitude that robotification could help solve. Many of these are ecological in nature. For example, even with our currently growing global population there are not enough human resources available to recover five continent-size ocean garbage gyres. Robots, however, fueled by sunshine, could very well help clean these up or at least keep them in check. Traditional agricultural machinery is becoming robotified, as autonomous platforms driven by real-time geospatial sensors. Such robots could just as easily help build and manage vertical farms, with substantially lower ecological impacts. And in the world of mobility, the dawn of autonomous electric vehicles and ride-sharing business models could radically reduce the amount of infrastructure required for human mobility, thereby reducing the ecological impact of modern human transportation. Of course, flying autonomous taxis would not even require roads.

The robots are coming, and this wave of automation, properly directed, could end up being the antidote to the ecological destruction brought by the last wave of automation experienced over the past century or more. Some fear the unintended consequences of autonomous technologies running amok. This has been an enduring theme in political discourse for generations, as pointed out by Langdon Winner in his book *Autonomous Technology: Technics-Out-of-Control as a Theme in Political Thought*. Yes, robots as another form of technological automation could very well lead to a technologically induced apocalypse. As it turns out, however, it seems that the dawn of the robot age, with robots using artificial intelligence and engaging in machine learning, may actually be the transformation in the world of automation that helps humans live within Earth's ecological boundaries.

Harnessing the Fourth Industrialization Revolution

Our discussion of Smith, Malthus, and Ricardo pointed out the cultural milieu of the First Industrial Revolution, of which they were such an important part. Every major era (and even every trend) in the life of our economy has shaped and, in turn, been shaped by the thinking of observers, scholars, and critics. This is no less true for the new economic era that we appear to be entering, which Professor Klaus Schwab, founder and executive chairman of the World Economic Forum, has dubbed, in his eponymous book, "the Fourth Industrial Revolution." Although his idea incorporates the dawn of the robot age discussed above, it goes a few steps further, with yet another set of implications for how we reimagine economics in an era of de-growth.

In Schwab's formulation, the Fourth Industrial Revolution builds on the personal computers, the internet, and the information and communication technology of the Third Industrial Revolution in a way that embeds a wide range of new innovations within the actual social structure of our societies, and even our bodies. In his view, robotics will be matched with artificial intelligence, nanotechnology, quantum computing, the Internet of Things, additive manufacturing and 3D printing, autonomous vehicles, and biotechnology in a kind of convergence that is actually hard to visualize, particularly given our basic assumptions about how industry and technology have shaped humanity historically.

Schwab views the first three industrial revolutions as driven predominantly by advances in technology. The First Industrial Revolution saw the rise of the factory, as a bespoke building designed to house the machinery needed to produce a specific manufactured good. The factories housing the mechanical looms for producing textiles are the archetype that we often harken back to, particularly since the organized theft of their Old World designs served as the foundation for the American Industrial Revolution. Yet the ironworks, the steam engine, and many more innovations were part and parcel of Europe's first sustained foray into industrialization. The Second Industrial Revolution, which spanned from 1870 to just before World War I, saw the continued growth of industries from the earlier era, along with the introduction of steel, oil, electricity, electrically powered mass production, visual and audio recording, telecommunication, the internal combustion engine, chemical engineering, and so much more. The Third Industrial Revolution again continued the growth of industries from both previous eras, but supercharged them with the fruits of the Digital Revolution. Despite all of these amazing innovations and the profound changes they brought to humanity, Schwab sees this Fourth Industrial Revolution that we are just now embarking upon as being woven into our social interactions and our very biology in a fundamentally new way.

Schwab envisions a world where every human and every human-made "thing," whether electromechanical or biological in nature, will soon become connected (or at least connectible) to an internet that is super-powered by artificial intelligence and machine learning on quantum computational infrastructure that is many orders of magnitude more powerful than we can conceive of today. These things will be able to be infinitely small and autonomous, and we will be able to produce them on command by 3D printing and other additive manufacturing processes.

Such a new era of connectedness and computation, embedded into our very beings and social structures, sounds both alluring and frightening. There is great opportunity to improve the efficiency of business and government, for better asset management, as well as the opportunity to achieve levels of sustainability that might create room for the regeneration of the natural environment that humanity has long burdened. Yet if the human-made material world and biological world do converge with the intelligence that Schwab suggests, operating at machine speed and at nano-scales, it does not take much of an imagination to think of a long list of catastrophic ecological outcomes that would become much more probable. It is particularly troubling to think of how this Fourth Industrial Revolution may turn out if these technologies are harnessed in pursuit of relentless economic growth, regardless of how well distributed its fruits are socioeconomically, as Schwab's World Economic Forum has pursued for decades.

Universal Basic Income

No serious discussion of automation and disruptive innovation can be had these days without some discussion of the concept of universal basic income. After all, if all this robotification ends up taking all of our jobs away, how would we possibly be able to afford to live without some basic form of income? Well, as it turns out, this concept is the subject of substantial debate.

A universal basic income, also called a "basic income guarantee," is a kind of social safety net concept in which all citizens of a country receive a regular, livable, unconditional sum of money, at a regular interval, from the government. Recipients are not required to work or even look for work, and the payment is given independent of any other income. There are, no doubt, other formulations or variants of this idea.

There is a certain attraction to this concept among the "small is beautiful" crowd, which sees capitalism as a kind of racket that needlessly drives acquisitiveness and wasteful consumerism. There is also an obvious attraction to this concept among those who place great value on the social wellbeing of those less fortunate, and who see people needlessly chasing the basics of subsistence rather than investing in achieving important personal and collective goals for the future. Many capitalists see "moral hazard" in concepts such as universal basic income, whereby recipients could see no reason to engage in productive endeavors, instead choosing to free-ride on others' economic accomplishments. Theory and practice, as we all well know, do not always see eye to eye. In this case, social experiments are the only path to useful knowledge. Some have already taken place, with mixed results. More, and different experiments with similar goals will emerge.

Where this fits in a world of de-growth remains to be seen. If a world of de-growth means that we will be continuously short of younger humans to serve aging humans, then it would seem to be a world of full employment, if only for geriatric services. However, we all know that a retrenchment in global population would lead to employment dynamics that are much more complicated than that. If anything, continuous declines in the number of working-age people paired with a rise in robotification, along with other waves of creative destruction, could lead to all manner of discontinuities in how our local, national, and global labor markets and economies evolve.

In the end, economists and their practical capitalist counterparts will need to forge a new set of economic concepts for the age of de-growth that we are about to enter. Whether universal basic income will be one of those core concepts is yet to be seen. Some new manner of social safety net surely will be. Regardless, when our obsession with "economic growth" (in the traditional sense) comes to an end, economic concepts that address the wellbeing of each person should be no less important than they were in the social welfare economics discussions of the 20th century.

CHAPTER ELEVEN

The Geopolitical Implications of Population Decrease

As we see around the world, various nations are already grappling with a decrease in population. There will always be domestic politics around this issue. The basic decennial reapportionment of representative democracy using a census will guarantee some level of tumult in those countries that enjoy such a system of governance. And there will always be resource implications when older generations require the financial support of smaller, younger generations. But what would be the geopolitical implications of worldwide population decrease? There are so many levels to this issue, but a few glimpses are instructive.

Beyond the softening of fossil fuel commodity prices associated with the rise of renewable energy sources, continuous decline in demand due to population decrease would have huge economic implications for nations that depend on oil revenues. There is a much more complicated energy transition already under way, putting long-term pressure on fossil fuel prices. Population decrease and the resulting decrease in consumption will only place additional downward pressure on prices. For those nations in which oil revenues are the only thing helping them deal with continuous population growth, how will that play out? And as the United States and European countries wean themselves off of fossil fuels, due to both renewable energy growth and population decrease, their geostrategic interests in the Near East will change materially.

For other nations, population growth and ethno-nationalism have been inextricably tied—even though it has often been more of an unreflective assumption than an explicit strategy. Hindu nationalists of the 20th century, when seeking to counterbalance or outnumber Indian Muslims and their Pakistani Muslim neighbors, were not displeased by runaway population growth, as though it would be a solution to communal violence. And India collectively seemed more than comfortable to have aggressive population growth that matched that of its northern rival China, as though masses of undernourished, uneducated peasants would win some future war for India. China seemed content to see its population grow, as living standards steadily rose from 1950 to 1980, until it became clear that its population was going to outstrip available resources. Similarly, over the past few decades, many Muslim-majority nations have seemed set on outnumbering their regional rivals, with explosive population growth rates and the resulting young populations. Some, like Iran, have even recently changed their family planning policies to induce population growth.

In truth, however, these are all examples of traditional cultures that have embraced many modern components of progress that enabled lower infant mortality, lower maternal mortality, improved public health, and increases in overall longevity while never evolving their prevailing, institutionally supported, cultural views on fertility—and more specifically, women's empowerment.[75] Perhaps in the haze of the 20th-century Cold War superpower competition, population growth was indeed considered the key to dominance and regional security. Yet hindsight seems to indicate that this was driven more by the half-

---

[75] To be fair, Iran's more recent pro-natalist policy is a retrenchment after decades of a very progressive policy of women's education, empowerment, and access to family planning technologies.

blindness that 20th-century cultures often had toward modernity—embracing the progress while neglecting to evolve their views on fertility.

Yet it would seem that at this point it is more complicated than simply working with these nations to adapt their views on fertility. Nations now face a sort of prisoner's dilemma in which their choices to rein in fertility will also be choices to embrace less economic growth and to forgo resources that may be of use in global geostrategic competition. What if China succeeded in reducing its population from approximately 1.4 billion to 500 million while India allowed its nearly 1.4 billion population to grow to 2.5 billion? What if India brought its population down to 500 million while other surrounding Muslim-majority nations refused to evolve past the explicit procreation guidance of the Koran and the hadith, so that Pakistan and Bangladesh moved from the current collective population of 350 million to 700 million? Or perhaps more acutely, what if Israel led the way to the new population plateau by reducing its population by 50%, bringing it from 8.5 million to 4.25 million, while all the surrounding Muslim-majority populations exploded?

Although these kinds of conversations make many blush, any transition from unmanaged runaway population growth to a thoughtful management of population strategies will quickly bring such issues to the fore. And if humanity cannot manage to overcome its tribalisms, if our many cultures cannot stare down their likely common disastrous ecological fate, and if we all cannot embrace the set of fertility norms that are required to escape global collapse, then there will be much more awkward and painful conversations ahead than these.

Let us begin this discussion by drawing some geostrategic distinctions between countries that have contributed massive population growth, or widespread industrialization of their national (and the global) landscape, or large volumes of persistent wastes in particular ways.

Demand Nations and the Geopolitical Pressures of De-growth

Particular nations have exhibited voracious demand for goods and services over the past century or more, and this demand has fueled much of the industrialization of the global landscape, and the creation of massive volumes of persistent wastes. In debates about who is responsible for global warming, many in the developing world point to the United States and all of its historical carbon emissions. In truth, its contributions to widespread global ecological destruction are much more widespread. Yet to follow this logic, many other developed nations—the nations in and around the European Union, as well as China, India, Russia, Japan, South Korea, Canada, and the like—all hold disproportionate responsibility for the ecological destruction of the past century because of their production and consumption. With the rise of the global middle class, we stand to see more and more countries becoming demand nations, with similar additional ecological consequences. As pointed out by the efforts of Wackernagel's Global Footprint Network, most nations are ecological debtors.

Thus, in any strategy that has global civilization moving to a new lower population plateau de-growth will be most acutely felt by these nations. It would require a profound exercise in self-sacrifice by these decidedly more powerful nations in order to make such a transition possible. As pointed out above, this de-growth does not mean a decline in personal wealth and wellbeing, if properly managed. Indeed, there is good reason to think that this path may be the only available path to achieving continued economic prosperity.

Key Ecoregion Stewards and Forgone Geopolitical Power

Whether they like it or not, certain nations must bear the historical and geographical burden of being stewards of key ecoregions. Brazil has disproportionate responsibility for the fate of the Amazon rainforest.

Indonesia has dominion over a massive collection of rainforests spread across 17,508 islands. The Congo Basin, the world's second largest rainforest, is spread across six countries—Cameroon, the Central African Republic, the Democratic Republic of the Congo, the Republic of the Congo, Equatorial Guinea, and Gabon. No doubt, many in these countries would like to harness these natural resources to their economic advancement, regardless of the ecological consequences. However, the global community has found itself exhorting these and other nations to preserve the integrity of the remaining parts of these key ecoregions, hoping that alternative economic models, such as ones based on ecotourism, can provide the necessary growth.

If Earth is to sustain the ecosystem goods and services needed for our own survival, key ecoregions must be protected, and even rewilded. Some ecoregions have been ravaged, perhaps to the point of no return, to hold enormous populations and critical industrial capabilities that people all over the world will depend on for generations. Although there may be opportunities to be more efficient in our use of these heavily burdened ecoregions—opportunities that we have a moral obligation to pursue—the truth is that we must pay disproportionate attention to a core set of very large ecoregions like the Amazon, the Congo Basin, the rainforests of Southeast Asia and Oceania, Arctic refuges, and key freshwater, estuarial, and ocean ecosystems. As of today, these key ecoregions still function and not all have hit critical tipping points. But, if they continue to be harnessed as part of global supply chains, their long-term ecological viability comes into question.

Watershed Hoarders and Water Wars

There are countries whose growth has demanded more water, and countries whose geographic advantages as headwaters nations have enabled them to control the flow of water to their downstream neighbors. Ethiopia's growing population, its thirst for energy and water, and its strategic placement on the Nile have meant disadvantaged water access for

Sudan and Egypt. Control of the Himalayan headwaters puts the enormous Indian and Bangladeshi populations at risk. The states of the American West and especially California (the latter sometimes aspiring to be its own nation) have worked to distort federal water management policies in ways that have enslaved upstream ecosystems through which ancient rivers have flowed for eons only to needlessly waste these precious water resources in horrific ways.

Watershed hoarding is a geopolitical phenomenon that will only worsen as humanity's ecological burden on our planet grows. As we unwind the population challenge, deliberate choices will need to be made about how to allot watershed resources and these processes will not be pretty—though, as we bring population down to a new, more sustainable plateau, questions of water rights will abate.

Some point to desalination as a panacea, using solar power to remove the salt from our oceans' water. The important question is, Where will the salt go? Already, Middle East nations have been dumping salt back into the Persian Gulf and other bodies of water, fundamentally changing their chemistry and long-term ecological viability. And salt is not the only issue. The complex cocktail of chemicals that are used to condition this water themselves accumulate as ecological hazards. No, the ocean will not be humanity's "Get Out of Jail Free" card, when the water wars come.

Facing the Impending Sub-Saharan Geopolitical Predicament

Making the conversation even more uncomfortable, we must call out specific geographies with population forecasts that place our planet in ecological peril. According to the United Nations Department of Economic and Social Affairs' updated medium-fertility scenario released in 2013, global population will rise from just over 7 billion in 2012 to nearly 9.6 billion by 2050—and roughly half of this growth will occur in the Sub-Saharan countries of Africa.

Much of the remaining population growth will occur in Asia, but for very different reasons, while most of the world's other regions are close to achieving replacement-level fertility or even dipping below it by 2050. Asia's contribution to this unsustainable 9.6 billion projection is largely due to the projected minimal incremental growth atop its already enormous population. There is actually a likelihood that Asia's population growth rate will decline below replacement level by 2050. North Africa, Latin America, and Oceania, all of which currently have fertility rates just above replacement level, are all projected to decline to slightly below replacement level by 2050. As mentioned earlier, North America and Europe are already below replacement level and are projected to remain there through 2050.

As the sole exception to this trend toward breakeven or below replacement level fertility, Sub-Saharan Africa leaves us in a bit of a geostrategic predicament, though it is not one born of great state competition, like the India-China rivalry. Instead, Sub-Saharan Africa has transformed into a target for both humanitarian investment and capitalist investment. Humanitarian investment by nation states and civil society organizations such as the Bill and Melinda Gates Foundation is providing Africa a much-needed respite from chronic illnesses, infectious diseases, and languishing wellbeing due to a lack of institutions such as schools and universities, and a lack of key government functions. Commercial entities and modern mercantilist governments such as that of China are investing heavily to lock in first-mover advantages, securing access to key geographies with key resources and key markets.

This investment will lay the groundwork for a growing African middle class. It will also fuel the massive projected Sub-Saharan population explosion if the same imbalance experienced in the 20th century is repeated. This should not be treated as an inevitability. However, its inevitability seems assured if the same geostrategic players who are currently poised to "help" Sub-Saharan Africa do not think about and invest proactively in providing for women's empowerment, women's education, women's workforce integration, and women's access to family planning technologies.

Helping Asia Learn from Japan and South Korea

As noted earlier, Asia is projected to be less of a net contributor to population growth by 2050 than is Sub-Saharan Africa, but, its contribution will still be massive because its current population is so large. Even a very small positive TFR on such a large base can be crushing. As such, it is important to point out that there is nothing inherent in Asian population dynamics that requires this population projection to be realized. All of Asia can and should reflect upon the experiences of Japan and South Korea over the past half-century.

Japan has actually been an industrialized power since the end of the 19th century. So, its path to negative fertility as a modern industrialized power should not be surprising. Japan worked tirelessly to rebuild after the devastation of World War II, and women were essential to these efforts. Japan can be cited for many failures in gender equity, like many other developed and developing nations; nonetheless, its empowerment, education, and integration of women saw their total fertility rate fall from 2.75 births per woman in 1950-1955 to 2.08 in 1955-1960. It remained at near-replacement level from 1960 to 1975, then fell slowly, reaching 1.49 births in 1990-1995.[76]

---

[76] The Population Division of the United Nations' Department of Economic and Social Affairs maintains country profiles, such as the one on Japan from which these statistics were drawn. Each of these country profiles offers fascinating insights into the respective nation's history. Reading these reports helps one understand the variety of national experiences with fertility, and how one should not simply see nothing but a juggernaut of inevitability in our planet's aggregate population trends. The Population Research Bureau is another great source for such national population profiles and broader trends, such as those here on South Korea. Their *Population Bulletins* distill complex demographic data and social science research to provide objective and accurate population information in an accessible format that anyone can understand.

Of course, over the second half of the 20th century, Japan enjoyed the benefits of progress with life expectancy at birth, for men and women combined, increasing markedly, from 63.9 years in 1950-1955 to 79.5 years in 1990-1995. This led to a marked aging in the Japanese population, with the median age rising from 22.3 years old in 1950 to 39.7 in 1995. For some, this trend was the source of much hand-wringing.

In contrast, although South Korea's late 20th-century experience is similar to Japan's, its starting point was quite different. Following the devastation of the Korean War, South Korea's population remained primarily rural and agricultural in the early 1950s. Its TFR exceeded 6 births per woman. In 1962, South Korea began a national family planning campaign by providing basic family planning information, basic maternal and child health services, and family planning supplies and services. The Koreans rightly saw this program as essential to the achievement of their goals of economic growth and modernization.

By 1970, the TFR had fallen to 4.5 and it was down to 1.74 in 1984. By 2005, it reached a historic global low of 1.08. This naturally led to robust discussion about low fertility, an aging population, and the introduction of pro-natalist policies to help buoy the nation's population growth. No doubt, this discussion was also driven by concerns over comparative economic growth statistics, with a focus on GDP growth rather than individual wealth and wellbeing indicators.

Although it is easy to reject such comparisons based on cultural differences or differences in economic and institutional histories, I would argue that national pride should be set aside in the face of the ecological calamity that awaits us. Asian nations have found common cause on many fronts, learning from each other and emulating each others' successful economic strategies. They have done well to collaborate in the construction of regional security alliances. I would suggest that they quickly huddle and trade success strategies for achieving negative fertility before explosive population growth exacerbates the low-intensity resource wars that are already afoot across Asia.

Protecting the High Seas from the Geostrategists

Humanity has spent millennia mastering the high seas for colonization, trade, transport, conflict, and resource exploitation. In so doing, and in spewing our wastes generated on land into the sea, we have managed to place our most important ecological resource in immediate peril. In the meantime, we continue to engage in geostrategic conflict, more through so-called "lawfare" than "warfare," to ensure military primacy in critical sea-lanes—rather than focusing on the oceans as a critical ecological resource needed to sustain human life on Earth. It is a curious "forest for the trees" kind of myopia to which it is easy to succumb in the face of the bravado of military and economic geostrategists.

The ecological devastation that modern industrialized humanity has unleashed on the world's oceans is nearing the farcical. Some 100 million sharks killed each year for fins to put in soup. Whales hunted to the brink of extinction, under government subsidy, for their meat, though it is largely disliked. Five continent-size garbage gyres unleashed on the wildlife and food chains of our high seas. Hundreds of hypoxic low-oxygen "dead zones" that are growing every year. Massive reef die-offs due to changing ocean temperatures and acidification. Invasive ocean species being transported inadvertently around the world, causing billions of dollars in devastation. And biogeochemical pressures being put on the very phytoplankton that generates some 80% of Earth's oxygen.

In the face of all this, we do not blink an eye at the increases in ocean cargo traffic that is spoiling key sea-lanes and the key coastal ecosystems around the world's 4,500 ports—and that will only grow further as the human population grows by 2 or more billion in the coming decades. And we forget that this increase will be far beyond linear, as these billions of people transcend poverty and join the global middle class, with vastly more voracious consumption patterns. It will come with more bilge water spoilage. Unintentional, frequent (and growing) numbers of fuel spills. Not to mention oil platform failures and tanker spills of devastating proportions.[77]

The increase in human population and human prosperity means an increase in industrial activities that will negatively affect our oceans, with massive geopolitical consequences. Nonetheless, it is entirely unclear that the global "geostrategic class" will elevate these considerations to live alongside military and economic discussions in the global geostrategic dialogue.

Recognizing the Middle East and North Africa as an Ecological Wasteland Requiring Geostrategic Investment

There are still some gorgeous and supple ecological zones across the Middle East and North Africa. But they are few and far between after millennia of humans engineering every floodplain for agriculture, cutting down every tree for fuel, shelter, and fiber, and introducing animal husbandry, which ensured the permanent denuding of the land. The Middle East and much of North Africa are an ecological wasteland, as humans have wantonly taken what their growing populations required, regardless of the natural growth cycles necessary to regenerate the historical ecosystems.

This has meant and will increasingly mean that this region will be inhospitable to the scale of human civilization that has developed there over time. It will mean a steady climb in the frequency and scale of water conflicts, famine and food conflicts, and energy insecurity across this region. Persistent low-intensity conflicts and more frequent large-scale

---

[77] This is not even to mention the fact that cargo ships are, collectively, tremendous carbon emitters, in 2018 burning some 300 million tons of very dirty fuel, producing nearly 3% of the world's carbon dioxide emissions.

conflicts will emerge due to the ecological frailty of the region in the fa
of these population dynamics. Yes, there will be beacons of moderni
such as the cities built by the emirates, and the like. However, these citi
will not sustain the quantity of human life that currently exists ther
which is only projected to increase. The ecological destruction that su
cities necessarily entail when built in such ecologically frail environs
hard to miss.

If we are to avoid consistent and contagious conflict in the Middle Ea
and North Africa, significant geostrategic investment must be made
cultivating ecological resources while lessening the human ecologic
burden on the region. This again means a robust cocktail of womer
empowerment, women's education, women's integration into th
workforce, and women's access to family planning technologie
Obviously, this proposal will not be immediately embraced by eve
regime across the region, given the reticence of many Muslim subgroups
empower women. Nevertheless, regional leaders must be engaged on th
issue, as many populations have simply outstripped the ecologic
resources of the geographies that they occupy.

There is also the need for ecological investments that could he
overcome the geostrategic fragility of the region. Efforts akin to the Gree
Belt Movement, begun by Wangari Maathai, for which she won the Nob
Peace Prize in 2004, offer the opportunity to rebuild water resources whi
reconstructing historical microclimate dynamics over the region. Efforts
rebuild wetlands and critical coastal zones offer the opportunity to rebui
important sources of food for local inhabitants.

There is no doubt that depopulating and ecologically rejuvenating th
Middle East and North Africa would have tremendous impacts on th
geopolitical stability of the region, resulting in a vastly increased quality
life for all involved.

******

Population decrease has many geopolitical implications, all of which are arrayed geographically in different ways. Some make us ponder what would need to be done to successfully navigate the process of population de-growth. Others are actually opportunities, where thoughtfully managing population decrease over certain geographies could help unwind specific gnarly problems. Still others indicate how our current geopolitical habits and narratives will likely make us fail to deal with the process of population decrease, subjecting large portions of the global population to unnecessary risk and degradation. It is a near inevitability that population will decrease by 2100. Yet assuming we manage to find our way to the glide path to population decrease sooner, before facing ecological disaster, these and other geopolitical discussions will need to become global dialogues, despite their unpalatable character.

CHAPTER TWELVE

A Cookbook for
Global Leaders and Global Citizens

It is easy to make predictions of impending doom. However, in truth, humans are moral beings with choices to make every day. As a species, we organize communities and governments in order to engage in collective action after we have come together to hammer out our differences and find common ground. As citizens and leaders, we can forge new pathways for our societies that help us avoid less desirable alternative futures. Humans, as a species, also have individual visions of the future and seek the liberty to pursue them by taking initiative and creating new enterprises. As such, there are useful discussions to be had about how citizens and leaders, together, can help transition to a new, lower population plateau of 3 billion—all the while inspiring individual citizens to act as entrepreneurs committed to this new path.

Nonetheless, since humans are prone to distraction, it is useful to have a cookbook of sorts to guide us all through what will no doubt be a complicated undertaking, if only to keep us all on the same page. If we execute these recipes with diligence and coordination, through both individual initiative and collective action, we will be able to avert catastrophe and build a resilient planet and a resilient human society that can support each other over the long term.

1. Establish a global campaign to empower women through education, employment, and access to family planning technologies. This campaign would have a major focus on removing the social and cultural coercion that so many women in so many cultures face to bear children by some arbitrary age and to continue bearing children without equal say in the decision. Call it the P3B campaign, as these new species-wide norms would quickly induce the below replacement level fertility required in order to help our species and our global civilization reach a new, lower population plateau of 3 billion.

2. Accelerate efforts to build sustainable, smart, and resilient cities, so that a large swath of humanity can withdraw from the core ecoregions that are needed to generate the ecosystem goods and services required to sustain the remaining 3 billion people, and move to cities without generating ecological debt.

3. Develop restoration and rewilding strategies for each ecoregion. To the greatest extent possible, as population is rolled back to this lower plateau, these strategies will be adapted to support advancements in rolling back the industrialization of our planet's landscape as well as its production and accumulation of persistent wastes. This would include a data-driven effort to identify geographies that require protection and formally establish further protected areas under the IUCN's framework, with diligence toward identifying and mitigating hollow conservation efforts by various regimes, in order to nurture the focal regions back to full capacity in their ability to

generate required ecosystem goods and services.

4. Begin systematically identifying the primary forms of human land use, particularly those in key ecoregions, and develop technologies and strategies to de-industrialize ecologically important parts of the global landscape, reducing humanity's overall ecological footprint. In particular, that means driving efficiencies and ecological sustainability into local, regional, and global food supply chains.

5. Drive wastes out of capitalist supply chains—both persistent wastes and the widespread application of non-persistent wastes, and recapitalize production lines to eliminate these wastes. This includes all forms of exhaust, including carbon, to the greatest extent possible.

6. Demand the design of low-impact infrastructure that provides modern industrialized humanity the mobility, water, and power that we need while providing ecoregions the ability to rebound to something resembling their historical ecological capabilities.

7. Stop impeding and diverting the natural flow of water, whether by dams, levees, causeways, or the infilling of wetlands. And reconsider the ecological impacts of current desalination strategies.

8. Stop using "growth" language in economic strategies and economic development planning efforts. And invest time and energy into reimagining core economic assumptions that will be needed in order to successfully navigate an era of "de-growth."

9. Reframe the "geo-engineering" debate in light of the two- to three-

century-long, unwitting, and hapless geo-engineering project that industrialization has wrought, and focus it on how certain interventions might be made to unwind the most egregious ecological offenses while avoiding unintended second- and third-order effects.

10. Think critically about how the land rights model of modern capitalism has driven autonomous (or perhaps omnivorous) technology to denude the planet, and develop alternatives that may help us to strike a balance between human development and protection of the ecological resources on which humanity fundamentally depends.

In effect, this would lead to a kind of cookbook for global leaders and global citizens to help usher our species into an orderly, not particularly onerous process of population decrease that does not entail genocide, eugenics, and other unsavory affairs that are typically invoked whenever one suggests bringing the human population in balance with our planet's ecological realities.

For those readers convinced at this point that Planet Earth is facing a massive ecological debt that threatens humanity's long-term viability, I challenge you to become part of the solution. Concerted collective action is required, but there are no end of opportunities for individuals, through personal initiative, to transform how humanity engages the planet that sustains our species. For those of you committed to making the world a better place, it may be worth exploring these 10 recipes in more depth.

A Global P3B Campaign

Perhaps this is stating the obvious, but women and girls constitute nearly 50% of the human population. And currently, they constitute 100% of the humans who bear children. The empowerment of women and girls holds the power to bring the TFR below replacement value, and in turn the global population to a new plateau of 3 billion, over some reasonable period of time.

Of course, for the length of human history, many institutions and cultural norms have systematically disempowered women of childbearing age from managing their own reproductive destinies. Although the barriers that these institutions and cultural norms have put on women have been very real, it is critical to remember the enforcers, who have too often been men. As such, any global campaign to achieve a planet of 3 billion must not only engage individual women worldwide. It must also engage men at every level, in national governments, in intergovernmental NGOs such as the United Nations and World Bank, in local governments, and in each household. Indeed, men in each household can have far more impact as supporters of women's empowerment than even the national leaders to whom they look for direction.

To be successful, a "Global P3B Campaign" would require several constituent parts. But the basic building block will be arranging for access to family planning technologies for every woman of childbearing age. Even if every woman and girl were educated, empowered, and integrated into the workforce, they would still require access to technological countermeasures to use against any man who would satisfy his sexual desires or religious imperatives at their reproductive expense. Where nations have no moral or legal opposition to women's empowerment and have the resources to provide family planning technology, progress is possible. Where nations demonstrate legal opposition on the basis of moral or religious foundations, progress is not going to be so easy. Where nations have the official will to enable women to undertake their own family planning strategies, but lack the resources to enable a supply chain for family planning technologies within their geographic boundaries, it will be critical for the international community to come to their aid. Too often, the

United States' domestic political landscape has undermined America's support for enabling women in developing nations to manage their own fertility effectively. Perhaps global intergovernmental NGOs, along with their donor nations, could close this gap. The Bill and Melinda Gates Foundation has already committed resources to help enable 120 million women and girls to control their own reproductive destinies by 2020 through access to family planning technologies.

Yet the challenge of enabling women's reproductive rights around the world is much larger than the challenge of tackling infectious diseases. The numbers of people who need access to these technologies is massive. By definition, it will always be all girls and women of childbearing age, which at this moment in history is nearly 2 billion individuals. As such, it is simply misguided to depend on private foundations to somehow marshal the totality of the required financial resources. However, the math is not complicated. It is basic geographic accounting.

While this basic building block is being put in place globally, we must embrace the fact that it is already in place in many of the countries that threaten the largest population growth over the coming decades. Thus, in those countries, we are left to drive public education among women and girls, and the men in their vicinity, so that they are empowered to save our planet.

Make Every City Sustainable, Smart, and Resilient

Cities, no doubt, have the potential to be the most ecologically efficient mode of human development. Still, we have a long way to go before every city on Earth can be considered sustainable, smart, and resilient. So many cities are vast ecological wastelands. Even when the formal portion of a city benefits from sound engineering for potable water, for sewerage, and for wastewater, informal settlements in and around them can be ecological nightmares that affect both environmental and human health far beyond the cities' limits.

Luckily much groundwork has already been done on the conceptual and practical foundations for how we can build sustainable, smart, and resilient cities. Sustainable engineering techniques have obsolesced older, ecologically heavy-handed engineering techniques in many places, leading more cities to "engineer with nature." But there is far to go. The term "smart cities" is being used to describe urban areas that use sensors, the Internet of Things, and other digital technologies to help manage assets and resources efficiently. Whether for more effectively monitoring and managing transportation or power plants, water supply networks or waste management, smart city strategies help optimize the efficiency of city operations and services for citizens and the environment. There are many dimensions to such technological strategies, but their potential for contributing to a lighter ecological footprint is clear. Then there is resilience. Cities face many kinds of challenges, ranging from climate change, to fast-changing migrant populations, to pandemics and cyberattacks. They also face complex emergent interactions with their local ecoregions, whether it is in times of drought or flooding, or famine, or human-made ecological disasters such as oil spills. Resilience strategies help cities to rapidly adapt and transform in the face of such challenges, whether expected or unexpected. Programs such as the Rockefeller Foundation's "100 Resilient Cities" initiative seek to build the capacity of individuals, communities, institutions, businesses, and systems to deal with both chronic stresses and acute shocks.

Still, most cities on Earth are not sustainable, smart, or resilient. Indeed, many are ecological scourges, infrastructurally dumb, and brittle in the face of both everyday stresses and frequent disasters. Every city must accelerate its efforts to become sustainable, smart, and resilient, so that a large swath of humanity can be withdrawn from the core ecoregions needed to generate the ecosystem goods and services required to sustain 3 billion people and moved to cities without generating additional ecological debt. If urbanization is to truly enable humanity to lighten its ecological footprint on our planet, much still needs to be done.

Develop Governance Mechanisms that Give Each Ecoregion a Voice

Our planet's ecoregions have no voice. They have no vote. They have no rights. Worse, the humans who have a voice, a vote, and actual rights have done little, if anything, to establish governance mechanisms that might develop thoughtful strategies for the protection of these ecoregions, and more importantly their sound and proper functioning. Although the IUCN has helped serve as a mechanism for the identification of specific protected areas, neither it or any other international or national organization has sought to systematically develop governance mechanisms for the world's ecoregions. Rather, we merely allow local sovereign authorities, or even just private actors, to make decisions for local ecological resources—decisions that have global implications—without any collective thought as to the larger strategic implications the ecoregions may have for our planet and our species.

To the greatest extent possible, as we roll back our population to a new lower plateau of 3 billion, these ecoregion strategies must be adapted to support humanity's efforts to unwind the industrialization of the global landscape, as well as our species' production and accumulation of persistent wastes. This would include a data-driven effort to identify geographies that require protection and formally establish further protected areas under the IUCN's framework, with diligence in identifying and mitigating hollow conservation efforts by individual nations, in order to nurture their focus regions back to full capacity in their ability to generate required ecosystem goods and services.

Much work has been done on a global scale to identify restoration and rewilding opportunities, such as the Atlas of Forest and Landscape Restoration Opportunities, developed by the World Resources Institute.

That publication identified more than 2 billion hectares worldwide that offer opportunities for restoration. This is an area larger than South America, with lands largely in tropical and temperate areas. But in which ecoregions do these disproportionately reside? And what would be the costs and benefits to the local population, and the sovereign nations affected, if any of these restoration and rewilding opportunities were harnessed in a quest to rebuild our planet's long-term ecological viability?

There is no practical scenario in which these determinations could be made by the global NGOs and intergovernmental organizations that have the expertise to identify these global opportunities. Governance mechanisms that are local to these ecoregions but focused on the ecoregion as their first-order object of analysis must be developed before we can have any hope of successfully bringing our planet back to a sustainable path, even assuming our world's human population decreases dramatically. Though our planet has the ability to flex its ecological muscles to reclaim what was once taken by human development, this is not true for every human-inflicted ecological scar it has endured. The collective action of humans, driven by common understanding and common cause, will be required for our planet and our species to thrive and prosper at a population plateau of 3 billion.

Identify and Transform Wasteful Land Use and Water Use Patterns

Much of our global landscape and seascape is suffering the effects of decisions on land use, water use, and marine area use made over a century ago when very different technologies were available to us. Little has changed about those decisions over the past century, other than the vast increase in the number of people demanding goods and services from this mode of production. In the 21st century, new technologies and business models have become available that offer us opportunities to radically refactor and transform how we use these vast tracts of Earth's surface. The incomprehensibly large geographies dedicated to farm land and pastoral land, and related patterns of water use are not an inherent consequence of the size of our population. Efforts to reduce food waste and to modify certain cultures' food mix could enable us to reclaim a large proportion of the land under cultivation, the land used for grazing, and the water demands of both. Some see this technological opportunity as a reason to find comfort in a future population of 9 billion, 11 billion, or 13 billion, mostly due to their failure to factor the accumulated persistent wastes into their equation. However, the transformation and integration of previously transmogrified lands and seascapes back into their historical ecosystems will play a major role in making our planet of 3 billion sustainable over the long term.

For many crops, vertical farming techniques can generate some 30 times more food per hectare than traditional farming techniques, with water and nutrient requirements that are orders of magnitude less. This could allow the transformation of lands that humanity wrested from their natural ecosystems during the 19th and 20th centuries, in particular. Other crops will require significant breeding or genetic engineering before they could successfully adapt to such strategies. This alarms many who see the use of genetically modified organisms as a threat in and of itself. Concerns over the introduction of genetically engineered monocultures across our global landscape have a real foundation in science and should be taken seriously. Other concerns over corporate control of our civilizations' core genetic seed lines have parallel foundations in political economic thinking and should be appreciated for their prudence. Yet there is a real opportunity to both lighten our ecological footprint by orders of magnitude while also cultivating the kind of genetic diversity that many seek—though perhaps altered in some cases for hydroponic production.

Our civilization's meat production is somewhat more problematic and difficult to resolve in a way that might enable the reclamation and restoration of our planet's landscapes, seascapes, and the water resources on which they depend and which they shape. Some 60% of the world's agricultural land is grazing land, supporting roughly 360 million cattle and over 600 million sheep and goats. That accounts for 26% of the planet's terrestrial surface. These grazing animals supply only 10% of the world's production of beef and about 30% of the world's production of sheep and goat meat. These grazing animals are, in effect, the local food sources for an estimated 200 million people—half in arid areas and half in other climates—in many cases providing their only possible source of livelihood.

The other 90% of the world's beef production and 70% of the world's sheep and goat meat production occupies even more land. This is not to even mention our insatiable desire for chicken. At any moment in time, there are 19 billion chickens on Earth. There are 1.4 billion cattle—yes, more than a billion in addition to those free grazing the planet's surface. There are some 2 billion domesticated pigs, in addition to all of the pigs, hogs, and swine that have been intentionally let loose for hunting purposes—becoming highly destructive and invasive species. This is matched by 1 billion sheep and 450 million goats, nearly a billion more than those grazing. All of these animals not raised under grazing regimes actually take up far less land, when you calculate the area of the fence lines and buildings that hold them in. Yet this is an illusion. The hay, feed corn, and other crops that they are fed occupy massive tracts of industrially designed, monoculture croplands that stretch as far as the eye can see. Although these animal husbandry strategies are more economically efficient, more profitable, and perhaps even marginally more land efficient, they generate enormous volumes

---

[78] This is not even to mention the fact that cargo ships are, collectively, tremendous carbon emitters, in 2018 burning some 300 million tons of very dirty fuel, producing nearly 3% of the world's carbon dioxide emissions.

of wastes that have a long track record of spoiling groundwater and surface water alike, leading ultimately to the generation of hypoxic dead zones in our lakes, rivers, and oceans.[78]

In simple geographical accounting terms, our roughly 7.5 billion people currently share a planet with roughly 7.68 billion acres of arable land. This is a roughly even split, where each human gets an acre. While this may sound like a lot to some urbanites, the average human omnivore demands roughly 3 acres of land for the full life-cycle implications of the animals we eat. Vegans, in contrast, demand only one sixth of an acre. In short, the more meat that we eat, the fewer people our planet can feed.

This discussion of how humans have reorganized Earth's surface in order to generate enough meat for our decidedly modern appetites is not complete without a discussion of our freshwater and marine area use patterns. The 20th century will be known for humanity's voracious destruction of the world's fisheries. More specifically, the 20th century will be known for the destruction of our fisheries due to overfishing, plastic pollution, and anthropogenic climate disruption. But to stick with the theme of modern humanity's insatiable desire for animal flesh, we have to appreciate that entire fisheries such as the North Atlantic cod have been fished to the point of collapse. We also need to appreciate that in harvesting species that are plentiful, such as shrimp, nearly 80% of all fish caught in the process are discarded—amounting to some 27 million tons of bycatch every year. And whether caught for their fins, or as bycatch, shark populations have collapsed globally, with white-tip shark populations having declined 99% since 1950. This loss of an apex predator fundamentally changes the balance in these marine ecosystems.

As it relates to land use, water use, and marine area use, there are endless opportunities for more sustainable transformations in how humans use our planet's ecological resources. It will not only require a change in mindset. It will also require sustained geographical analysis of different forms of resource use, and research and development into more sustainable approaches.

Do not operate under the illusion that simply bringing the world population down to a cool 3 billion will make the world ecologically sustainable over the long term. The 3 billion estimate that I provide is that of a technological optimist and an unrepentant capitalist. When it comes to how we use our world's resources, innovators from government, industry, academe, and the social sector will all have to undertake a heavy lift in helping to transform how our species harnesses our planet for our own benefit. We will need to begin by systematically identifying the primary forms of land use, water use, and marine area use, particularly those in key ecoregions, and develop technologies and strategies to de-industrialize these parts of the global landscape, reducing humanity's overall ecological footprint. In particular, that means driving efficiencies and ecological sustainability into local, regional, and global food supply chains.

Eliminate the Wastes of Modern Industrial Supply Chains and Human Consumption

Many fans of capitalism are blind to its wastefulness, to the fact that capitalism prolifically generates persistent wastes that accumulate across our planet's ecosystems, as well as in our atmosphere. Capitalists, and the economists who have elevated their craft to a body of theory, focus instead on how capitalist economies allocate resources more efficiently than their socialist counterparts. But this use of the term "efficiency" should never be confused with the notion that capitalist economies do not generate enormous volumes of waste.

If we are to build a global society of 3 billion humans capable of sustaining itself over the long term without generating ecological debt, we must look seriously at how to eliminate as many wastes as possible from the existing capitalist supply chains. We must figure out how to eliminate the widespread and continuous generation of wastes, including all forms of "exhaust" such as carbon and so many other wastes and contaminants that we spew into our atmosphere at industrial scale. We must also find our way to generating less incidental waste, such as consumer packaging, and reducing the amount of resources expended to achieve our consumption goals. Of course, less consumption would also be ideal, but there are practical limits to how much less. After all, modern industrial humans, by definition, produce goods and services for consumption.

The old adage "reduce, reuse, recycle" is no less important to our future than it has ever been. Regardless of how diligently we follow this guidance, our remaining production and consumption will still generate wastes that could potentially be engineered out of existence given sufficient attention over the long term. It simply requires recognizing the "art of the possible."

Finland has decided that achieving zero wastes, including zero climate emissions, is possible. In taking the initiative to achieve this goal, they have shown us that, indeed, humanity could enjoy a high quality of life while vastly reducing the enormous volumes of waste that burden our planet. On average, each Finnish person generates 500 kilograms of waste per year, if we count only their household waste. If one thinks of this as the tip of an iceberg, about half of this iceberg comprises waste from extractive industries, and one fifth is industrial waste.

In the Finnish perspective, there are opportunities far beyond more efficient household waste management. They see opportunities for factories and plants to engage in industrial symbioses that make possible much more efficient exploitation of side waste streams that would otherwise spoil Earth as postindustrial waste. These opportunities, however, will often not be realized through the normal economic signaling of typical capitalistic economic interactions. Efficiently exploiting industrial waste appears to require some institutional coordination to unleash the possibility of creating new business opportunities for companies in particular, while raising productivity.

The Finns see this process as the creation of a circular economy in which most materials are renewable and those that are not are efficiently recycled and reused by industry in the manufacture of new products. They have not yet figured out how to deal with all complex heterogeneous waste materials such as alloys, textiles, and plastics, but their reuse rate is continuously rising.

The question becomes how to get producers to engage in ecologically friendly designs that mesh well with the state of the art of recycling, such that producers take responsibility for the wastes that their goods generate. However, without global coordination, under a single set of principles that can be applied on a global basis, the global supply chain from which many producers source their materials will always make this a challenge.

Another word on humanity's collective exhaust is due. As we drive wastes from our industrial supply chains, this effort must include a goal of achieving a carbon-neutral society. But carbon-neutral is not enough. We must also recognize the wide range of other wastes that we regularly vent into our atmosphere through, or associated with, various forms of combustion. Methane, nitrous oxides, and fluorinated gases also contribute to global warming—with much higher impacts than originally thought. And sulfur dioxide, mercury, and other toxic metals that enter the atmosphere through combustion directly affect human and ecosystem health, and accumulate in the global food web. It is crucial that we not only eliminate the wastes that we can see with our eyes, but also the wastes that change the very chemistry of the finite biosphere in which humanity and the ecosystems that support us are sealed for the foreseeable future.

Wastes can be driven from global capitalist economic supply chains, but not through capitalist economic interaction alone. Institutions that supply information, broker commercial synergy, and spark a wide variety of new innovations need to be brought into being if we are to have a population of 3 billion thrive and prosper without incurring long-term ecological debt through the accumulation of persistent wastes like those that dominate the global landscape and seascape of today. Without some sort of technological optimism, there is no way even 3 billion humans will be to live on Earth over the long run without smothering under the weight of accumulating ecological debt.

Transform Infrastructure to Have Net-Zero Ecological Impact

Infrastructure is a human innovation that serves as the basic physical and organizational building blocks for the operation of civilization over time. Modernity requires infrastructure, and indeed, our infrastructure has enabled many ecological savings that have made progress possible as our population has grown. Places that lack modern infrastructure demonstrate how growing human populations can become ecological nightmares in particular geographies. Without modern wastewater and sewage management infrastructure, human waste can spoil watersheds and coastal areas. Without modern transportation infrastructure, human mobility leads to widespread erosion. Without modern electrical power infrastructure, local populations resort to deforestation in order to meet their fuel needs.

Infrastructure strategies that work within one ecoregion often do not transfer well to other ecoregions. Tropical, desert, temperate, and cold ecoregions exert their own realities on human infrastructure innovations, and demand that humans adapt their infrastructure strategies to local conditions if they are to be successful. The transplantation of infrastructure techniques that work well in one ecoregion to another ecoregion often leads to adverse consequences for both humans and the environment. There are many historical cases of the misapplication of various infrastructure types to ecoregions and climates for which they are not suited, leading to ecological disaster. Take road infrastructure as an example. For roads to work equally well for drivers in different ecoregions, different road engineering techniques have had to be developed to adapt to local realities, whether to cope with extreme heat or cold, erosion and water drainage in the face of torrential downpours, or geological instabilities. For roads to work well for local species in different ecoregions, other adaptations are required—though most often, such adaptations are neglected, leading to millions of animal roadkill deaths annually.[79] Beyond roads, many infrastructure methods deemed very effective in one ecoregion have been applied to far-flung ecoregions with very different ecological sensitivities that do not exist in the infrastructure method's place of origin, magnifying the ecological burden in the ecoregion of new application. This has actually been an all too frequent outcome, as Western infrastructure techniques have spread around the world.

---

[79] Although I alluded to the ramifications of impassable mobility corridors earlier, the ecological impact of roads and cars specifically simply cannot be overstated. Roads that were constructed with little to no understanding of local ecological processes and animal movement patterns have led to a cataclysmic wave of roadkill deaths all over the world over the past century. This is hardly a new insight. In 1920, on the basis of his observations, the naturalist Joseph Grinnell estimated that thousands of animals were being killed in California every day due to automobiles. Beyond altering and isolating animal populations and their habitats, deterring the movement of wildlife in the regular conduct of their life cycles, roads and cars have also caused extensive wildlife mortality. Unfathomably large numbers of mammals, birds, reptiles, amphibians, and invertebrates are killed on the world's roads every day, with estimates at more than 1 million a day in the United States alone. Although some are inclined to dismiss this phenomenon as affecting only deer and other "nuisance species" that are experiencing runaway population growth due to the loss of apex predators in their ecosystems, the truth is much more complex. Many endangered species, such as Florida panthers, desert tortoises, koala, and a host of birds, face existential threats from roads. Disturbingly, not only are our winter road-salting operations leading to unhealthy rises of salinity in nearby surface and groundwater, but this salt also attracts animals in search of saltlicks, thus leading to even more roadkill deaths.

The question, then, is how citizens and decision makers can demand the design of low-impact infrastructure that provides modern industrialized humanity the mobility, water, and power that we need while providing ecoregions the ability to rebound to something resembling their historical ecological capabilities. In a world where the need for mobility solutions too often leads to the construction of more roads, we have developed ecologically impenetrable mobility corridors that have cordoned off enormous geographies such that the historical wildernesses could never rebound. In a world where humans demand more freshwater for themselves, their crops, and their domesticated animals, far-flung water infrastructure has diverted water resources from their historical natural flows toward our faucets and irrigation infrastructure. In a world where power needs have led to the construction of more plants burning fossil fuels, our atmospheric and ocean chemistries have been forever altered, at a breakneck pace, while hard metal wastes have rained down on local and distant geographies.

The 20th century saw massive global proliferation of infrastructure development, which enabled progress of many kinds. Toward the latter part of the century, it became clear that many of these infrastructure concepts and techniques simply lacked an appreciation of the ecosystems within which humans hoped to prosper, and how they might undermine the local ecosystem's long-term viability—thereby threatening local human wellbeing in specific locations all over the world. Some cities and regions engineered infrastructure that created an ecological safe space for themselves, while shunting the ecological burdens to others downstream.

Much work can and should be done to figure out how to undertake new infrastructure development to meet human needs while having net-zero ecological impacts. Needless to say, even more work can and should be done to transform much of the 20th-century infrastructure, which in retrospect was so clearly ecological folly, into far less impactful and more resilient enablers of a society of 3 billion prosperous individuals.

Revert to Our Planet's Natural Flow of Water

Humans have been diverting the natural flow of water for millennia. However, when only millions or hundreds of millions of people inhabited the planet, in an era that preceded large-scale, industrial-age, water management infrastructure, there was little in the way of ecological impact. Notable early exceptions were the national canals and irrigation works of early China, such as the Grand Canal and its predecessors as far back as 1890 BCE.

Early industrialization in Europe and in America also brought with it the diversion of water. But it was the 19th and 20th centuries that saw water diversion at epic scale. The building of hydroelectric dams, the damming of reservoirs, the creation of lakes for recreation, the digging of canals, the transformation of rivers into navigable waterways, and the hijacking of groundwater for irrigation and potable water were all undertaken at unprecedented scales and in unprecedented numbers. Then the United States and other developed nations exported these techniques to the rest of the world.

The 20th century's vast diversion of the natural flow of water will go down in the history of the world as a major misstep. Reverting the natural flow of water, if we can manage it across so many diverse geographies, would become one of the major accomplishments of the 21st century. Of course, existing populations demand these diverted waters and make it difficult to imagine reverting these flows until such populations begin to decrease.

Reimagine Economic Concepts to Prepare for a World of De-growth

Economics, as a body of knowledge, is relatively young. The economics of long-term economic growth is even younger. The work of Smith, Malthus, and Ricardo has been augmented by that of scholars seeking to understand what drives long-term economic growth. Alfred Marshall, Joseph Schumpeter, John Maynard Keynes, Robert Solow, Richard Nelson, and others helped us understand the processes driving economic growth over just the last century or so. Now, we need a new generation of scholars who can help reimagine the field of economics to prepare us for a world of de-growth.

Regardless of whether my estimate of 3 billion as the actual carrying capacity of the planet is correct or not, it is a near inevitability that by 2100 the world's population will peak and begin to decline, if ecological annihilation does not occur first. As such, it seems that for the next few decades, the field of economics should work diligently to help the practitioners that we call capitalists to build a prosperous and sustainable society in the face of population decrease.

I would suggest that the scholars mentioned here, who helped unlock our understanding of the economics of technological change, particularly Richard Nelson, should be looked to for guidance in building this new body of knowledge. The rate and direction of scientific advance and technological change can be shaped by investment in all manner of supporting institutions. These institutions, in turn, can help us realize the application of a wide range of values through our collective economic action. Properly crafted, these institutions and investments could help us fashion a future in which technological innovation, automation, productivity improvements, and growth in collective and individual wealth and wellbeing enable us to balance ecological resilience and human prosperity.

What remains clear is that continuing on the path of economic thought and policy focused on GDP growth will lead us over the cliff and leave us unprepared for the future that we face together.

Evolve Geo-Engineering from a Hapless, Inadvertent Process to a Deliberate, Strategic Process

Humans have been engaged in hapless, inadvertent geo-engineering for millennia. Whether it was the transformation of local microclimate dynamics through deforestation, or whether it was the lead spewed into the atmosphere by metal smelters for thousands of years, humans have been engaged in large-scale geo-engineering without ever even knowing it.

This clueless alteration of the very geochemical processes at the heart of our planet's carrying capacity only accelerated with industrialization. The industrialization of carbon emissions gave way to the globalization of a carbon-emitting civilization at a scale that has led to steady increases in mean atmospheric and sea surface temperature, albeit as the oceans absorbed so much carbon as to lead to a deadly acidification. Together, the heat and carbon already absorbed by the oceans will take centuries to dissipate, even if all industrial carbon emissions halted today.

Of course, focusing on industrial carbon emissions is a gross simplification. Methane emissions are far more impactful as greenhouse gases. Methane is emitted by gas flares, burning off natural gas at oil wells and other energy sites all over the world, as well as by leaks from natural gas systems.[80] It is also a natural by-product of the billions of head of livestock that we so unnaturally raise industrially on our planet. In 2018, many focus on the 1.4 billion cattle that pass methane in copious amounts, but the 2 billion pigs, 1 billion sheep, 0.5 billion goats, and 19 billion chickens also pass methane on a daily basis. Methane is also emitted

---

[80] Although flaring is meant to destroy methane in lieu of releasing it into the atmosphere, this is an imperfect process. Because flaring is considered a waste disposal process, there is no systematic reporting of the locations where flares occur or the gas volumes that are flared. It is estimated that some 143 billion cubic meters of gas were flared worldwide in 2012, with the number only increasing since then. A satellite remote sensing method called VIIRS, or Visible Infrared Imaging Radiometer Suite (commonly referred to as the "Lights at Night" data set), found more than 7,000 flare sites across the globe. Russia flared the largest volume, and the United States had the highest number of flares (2,399).

naturally from wetlands, but most disturbingly, as Arctic regions warm due to climate change, the massive methane reserves trapped below the tundra are poised for a catastrophic release. This component of humanity's unintentional geo-engineering effort should alarm everyone.

No one should recommend any large-scale geo-engineering interventions in a casual way. Humans have a very bad track record at anticipating second- and third-order effects of their actions. This is why we have the term "unanticipated consequences." Yet if everyone recognizes that we—modern industrial humanity—are already actively invested in an agglomeration of thoughtless, unintentional geo-engineering activities, then everyone might also be willing to entertain ways to alter our production and consumption behaviors to make these activities less harmful to our planet. Only then might we also engage in a more deliberate strategic discussion about geo-engineering interventions that might match the scale of the problems that we have accelerated over the past two centuries, to unwind our most egregious ecological offenses while avoiding those unintended second- and third-order effects that so often haunt us.

Reconceive the Modern Land Rights Model
Since ancient times, humans have created all manner of land rights models with which to divide ownership of given geographic domains. The ways in which land rights models have evolved over the centuries have had major implications for how the industrialization of our global landscape has unfurled.

The Age of Exploration, and the colonial period that it gave way to, instilled a kind of "finders keepers" ethos to the lands that Europeans "discovered" and conquered (or stole)—typically in the name of the sovereign whom they served. The communal property rights of indigenous peoples, lacking written deeds or titles, were simply ignored and swept away through use of force and duplicity, at the convenience of the Western powers that washed up upon their shores. There was no recognition on the part of these Westerners of the geographical ebbs and flows of the indigenous people's relationship with the land based on seasonal weather patterns, animal migrations, and biological productivity—and how this responsive relationship was reflected in indigenous concepts of land ownership.

Interestingly, in England, it was not until the Enclosure Movement in the 16th century that the crown's presumed ownership of the entire realm gave way to individual land rights. For at least a millennium beforehand, commoners farmed communal lands owned by the crown. The Enclosure Movement saw these communal lands fenced and deeded or titled to one or more owner. These owners then had economic incentive to "improve" the land in ways that could generate economic value for themselves, beyond mere subsistence farming. Other models, such as the cadaster implemented in Europe under the Napoleonic Code, which fundamentally reshaped land rights across the continent, had similar impacts upon how individual land owners interacted with the land.

In a completely different cultural context, beginning in the 17th century, imperial China also began a long process of transformation of its land rights model. Although scholars disagree on some aspects, it is generally understood that before this period, landlords paid taxes to the imperial government in return for land-owning rights (so-called subsoil rights), that allowed them to collect rent from peasants who used the land for agricultural purposes (so-called topsoil rights). Wars, rebellions, and natural disasters, of course, led to more complexity, frequent shifts in ownership, and the occasional abandonment of both subsoil and topsoil rights. But under the Qing Dynasty, beginning in the mid-17th century, some titling of land ownership did begin to take place—though it did not have the scale or economic impact of the Enclosure Movement in England. Under the 20th century Nationalist government (1912-1949), the Chinese came to embrace statutory land rights, grafted to German and Japanese civil law traditions, while retaining all manner of communal and customary rights vested in landlords, nobles, religious institutions, and village communities. None of this did much to unleash the power of capitalism the way the English Enclosure Movement did when paired with the First Industrial Revolution.

In the United States, it was something more akin to the English model that drove land rights—a model that was later re-exported to much of the world. Accelerated by state-sponsored homesteading policies, this model led to the rapid expansion of industrial techniques across a previously pristine wilderness (if we overlook the material but modest ecological burden the American indigenous people placed on the land).

In Hernando de Soto's many books, including *The Mystery of Capital: Why Capitalism Triumphs in the West and Fails Everywhere Else*, he makes the case for why this modern land rights model has enabled economic growth in ways other systems failed to do. Land administration systems that allow for the encoding of private land rights in deeds or titles create a legally enforceable abstraction that unleashes the wonders of modern capital. This is a system that enables people to legally present land as collateral in a system that enables people to take economic risks, which if they pan out, lead to profit and capital accumulation. In turn, this capital can be exchanged far and wide in economic markets. Many informal economies exist without such formal administrative land rights systems, leading to a sort of informal capital in the form of land, which is no less valuable but which is trapped. In these systems, the "owners" cannot unlock the power of this informal capital and use it to transform their lives.

The modern land rights model, while productive in unleashing the mysterious powers of economic capital, has also unleashed an unstoppable wave of ecological destruction wherever it has been implemented. The ability to risk and accumulate the capital associated with land has generated a voracious appetite for acquiring and "improving" land. Needless to say, such improvements too often involve the deletion of much of the historical ecological value, or natural capital of the land, and its replacement with physical capital improvements. While a boon to economic growth, it has been the bane of ecosystems around the world.[81]

The news is not all bad. As it turns out, capitalist landowners and ecological preservationists have found a way to use these modern land rights models to benefit the environment, where politicians have been willing to cede some ground in the tax laws. Together they have created "conservation easements," which are restrictions placed on a piece of property by its owners to protect its associated natural resources. These easements can be either sold by the landowner to a conservation organization or voluntarily donated. In either case, the easements end up legally preventing certain types of development and uses, in perpetuity. Such legal innovations can harness the power of private investment or private philanthropy to protect ecologically sensitive lands, and organizations such as the Nature Conservancy have used this mechanism to

---

[81] In his book *The Value of Species*, Edward L. McCord makes the important distinction between the power of Western land rights regimes and the power of money. Although owning its own property might enable a family to graze its livestock more productively, without money as a method for exchange, the benefits of land ownership, and thus the pressure the develop land, remain limited:

> If the sheep in our pasture are only to feed and clothe our family, then there is a limit to the number of sheep we can use. But if the sheep can make money for us, and we have access to a global market for them, then the number of sheep that we can use increases exponentially, and so does the pressure on the pasture. Indeed, our incentive to protect the carrying capacity may even diminish in some cases relative to the size of the short-term profits to be made and the option of taking them elsewhere. In effect, as money gains more purchasing power, in the global market economy, the conquest of things from the earth that make money expands indefinitely with market demand. (page 66)

Thus it is not just modern property rights regimes, but also modern monetary infrastructure, that have unleashed a seemingly unstoppable value proposition that has seen short-term economic calculations eclipse long-term ecological considerations.

great effect. Sadly, this model does not apply to the oceans.

All of this, however, focuses on the rights of humans, in a model that assumes human dominion over the natural resources of Earth. Nowhere in these approaches is there a conception that the planet preceded humanity, that humanity is born of our planet's wildernesses, or that these wildernesses deserve protection in their own right, since it is they that underpin the long-term ecological viability of our species in the first place. This worldview has raised the question of whether nature, or particular natural landscapes (or seascapes), should have their own land rights. Should ecoregions have a voice, in a legal sense? This has spawned an area of law that some refer to as "Earth law" or "rights of nature law."

This may sound ludicrous to those of us inculcated with a modern Western legalistic worldview, but to indigenous peoples whose very cultural existence is predicated on their ties to their ancient lands, this is not the case. For instance, after 140 years of negotiation, a Māori tribe won recognition for Whanganui River, on the north island of New Zealand, meaning that it must henceforth be treated as a living entity. After centuries in which indigenous people's rights to land have been questioned or simply rejected, because of the lack of a deed or title, perhaps such rulings will elevate the crux of the matter in the Western legal consciousness. Indigenous people have never claimed ownership of these natural resources because in general, in their belief systems the land itself is a living entity that deserves respect and protection from human depredation.

Advancements in the field of Earth law thus far are modest but are important to note, particularly as they relate to our oceans and rivers, which have historically been excluded from private land rights regimes and therefore protectable by conservation easements. An Earth Law Framework for Marine Protected Areas was introduced at the 4th International Marine Protected Areas Congress in 2017. A Universal Declaration of River Rights has also been developed. While far from having national-scale rights of nature laws on the books, advancements in both local ordinances and case law appear to show momentum for this worldview. More importantly, there appears to be a recognition that this giant hole in the Western land rights model has been exported globally, at least as it relates to the ecological wellbeing of oceans, rivers, and other bodies of water that are largely ignored in existing legal frameworks, outside of standard riparian water rights. In the end, whatever legal mechanisms we choose, we must think critically about how the land rights model of modern capitalism has driven technology-intensive development to denude large swaths of Earth, and we must conceive of creative alternatives that may help us to strike a balance between human development and the protection of the ecological resources on which humanity fundamentally depends.

CHAPTER THIRTEEN

The Consequences of Ignoring Reality

So what are the consequences of ignoring the reality that our planet can only support 3 billion modern industrialized humans?

Well, a world of 9 billion, 11 billion, or even 13 billion people is fast approaching. This will bring such widespread industrialization of the global landscape and such enormous volumes of persistent wastes that our terrestrial and marine ecosystems, as well as the atmosphere that they fundamentally shape, will become unrecognizable.

The ecological debt accumulated over the past century, in particular, not to mention the patterns and behaviors of the current 7.5-plus billion people we have on Earth, have unleashed a juggernaut of ecological destruction that shows no signs of slowing. Forests are falling at alarming rates, oceans are acidifying, reefs are dying, freshwater sources are becoming spoiled, innumerable species are facing collapse, large ocean ecosystems literally cannot breathe, climates are being altered, glaciers are disappearing, and sea-level rise is threatening coastal human settlements that constitute a massive proportion of the global human population and trillions of dollars in assets.

While political opponents face off on issues of human carbon emissions and the vagaries of atmospheric chemistry, neither side of this debate is paying attention to the geography of humanity and the punishing ecological footprint that a human species of our current size and geographic distribution has wrought on the planet. In the end, humans are the source of this ecological destruction, whether it is too much carbon, too much impervious surface, too much industrialized agriculture, too much diverted water, too much waste injected into our oceans, or any other of the lengthy list of human ecological impacts. This is all to say that although climate change is real and will have an enormous negative impact on our planet and our species, the totality of humanity's ecological debt is far, far worse because of the sheer number of humans on Earth. A population of 7.5 billion people is already fundamentally unsustainable, so through our fertility patterns we will be making a life or death decision for our planet—one that will lead us either to convulsively undermine its carrying capacity or to help mend the ecological wounds that Homo sapiens, as an invasive species, has caused.

Over a half-century ago, Rachel Carson brought to the world's attention the scourge of DDT, raising the possibility of waking up to a silent spring. Sadly, the fact that this was merely one of many endocrine disruptors has largely escaped the popular mind. In the following decades, many people woke up to the threat of human-induced climate change, and the role of the industrial emission of carbon and other greenhouse gases. Decades later, we still lack a global consensus on climate change, with corporate and sovereign interests shaping the denial of a small but politically potent segment of the global population. Even though steadily growing ocean dead zones were identified in the 1970s and have been monitored over the half-century since then, their existence has not yet entered popular discourse. The same for the continent-size—and growing—ocean garbage patches, which evaded detection until the mid-1990s, and remain little more than a curiosity two decades later. Modern, urban humans have been conditioned to think of bucolic and pastoral landscapes as natural landscapes, when they actually represent the permanent, mechanically and chemically induced denuding of the historical ecosystems that once dominated those same geographies. And whether urban, exurban, suburban, or sprawling informal settlements, the impervious landscapes that modern humans have paved across Earth's surface have all but eliminated the production of the historical ecosystem goods and services that those geographies once generated for our planet and our species.

As a species, *Homo sapiens* has demonstrated its willingness to look forward to the next accomplishment and to call it progress, while failing to monitor and account for the second- and third-order effects of those accomplishments in the aggregate and over time. Not only have we demonstrated a proclivity for ahistorical thinking, our tendency toward ageographical thinking has made it difficult for us as a species to understand the implications for our future on this finite planet.

The implications for the future are dire. It would be quite easy for humanity to simply behave like the proverbial frog in the pot of water. Just as the frog would jump out suddenly if placed into boiling water, humans born into our historical ecological habitats might be alarmed if they were abruptly thrust into our current situation. However, born into the modern industrial landscape, we who are presently entrusted with the fate of our planet are more like the frog placed in tepid water, which is brought to a boil slowly.

The prognostications at the core of this book may seem absurdly dire to some. If you agree with my calculations of the limited carrying capacity of our planet, I implore you to think of yourself as a frog being slowly boiled by the decisions of previous generations and your own patterns of behavior that you have been born into.

*****

What happens if, indeed, the ecological debt we have incurred in the past century or so is simply too great, and we can't buy it back down in time? How should we think about this situation? It is not just that sea level will rise, island nations will disappear, coastal cities will recede, ecosystems will collapse, ecosystem goods and services that sustain humanity will be eliminated, and we will magically remain at the optimistic 9 billion population peak posited by the UN. We would have to make incredibly drastic plans or simply cut bait and let an ocean of people (no pun intended) fall into the abyss, and somehow keep the privileged minority protected from the ensuing onslaught of global pandemics of shocking proportion, in a world where mobility and connectivity—connectography—make such buffers unimaginable.

Thus, a planet of 3 billion should be our goal.

An Epilogue on Planetary Ethics

Too many expert analyses of our impending future seem to emphasize macro-level trend lines that need to shift if we are to avert doom, while wistfully gliding over the ethical issues involved. Our recent global exercise of generating Sustainable Development Goals (SDGs) strikes me as another one of those. Although this exercise was a critically important process that laid the groundwork for collective action on climate change and other ecological issues essential to our planet's long-term sustainability, it was a political process that was long on posturing and short on making people think about their own ethical obligations to each other, and to the planet that supports us as a species.

The SDGs, while ambitious, are a complex set of political compromises, as any such effort inherently will be. Although informed by expert analysis, the process resulted in a list of inconsistent goals that are difficult to quantify, implement, and monitor. In short, there is a gaping inconsistency between the socioeconomic development goals and the environmental sustainability goals. Although many have focused on the difficulty of measuring and monitoring progress on (or defection from) this set of non-binding goals, at least there is some expectation that each country will create its own national and regional plans.

Unfortunately, the SDG process considered population and population growth dynamics as exogenous and inevitable. There was no attempt to calculate the carrying capacity of the planet at the outset of the SDG process. As a result, there was no real discussion about the efforts that nations and citizens would need to undertake in order to taper populations in order to reduce their ecological burden to a level that would allow our planet to thrive over the long run. There was no quantitative, geographic accounting of the harm that each country, each square kilometer of people, and even each person is doing to the planet. And without a notion of harm, it is difficult to have a discussion about ethics.

In the field of ethics, the "harm principle" says people should be free to act however they wish unless their actions cause harm to somebody else. The harm principle is a central tenet of the political philosophy known as liberalism and was first proposed by English philosopher John Stuart Mill in 1859, in his book *On Liberty*. Mill first articulated this principle in these words: "The only purpose for which power can be rightfully exercised over any member of a civilized community, against his will, is to prevent harm to others"—echoing a similar articulation in France's "Declaration of the Rights of Man and of the Citizen" of 1789.

If you agree that we have exceeded the actual carrying capacity of our planet, each additional human on Earth is, by definition, doing ecological harm to the planet that cannot be avoided, due to the ecological debt that previous generations have already accumulated.

At its heart, this is a question of ethics. Different countries, and the cultures that make up their societies, have managed to find themselves at different points on the population growth continuum. Frankly, most nations' positions on this continuum are a result of chance and drift, not purposeful policy outcomes. The nations that happened to reach a certain level of economic maturity and happened to embrace women's empowerment and integration into the workforce also happened to achieve something close to replacement-value fertility, or even negative fertility—often to their own chagrin. Most countries with runaway fertility did not choose this growth, though they may have unwittingly chosen to embrace behaviors that in the end fueled it.

If leaders and citizens around the world know the facts about fertility and the ecological limits of our planet, our dilemma then becomes an issue of ethics. It becomes an issue of informed personal choice, as long as nations and localities do not constrain personal choice—for instance, with barriers to access (either legal restrictions or supply chain limitations) to family planning technologies.

Any discussion of what part of Earth, or which nation, is overpopulated, quickly becomes a discussion of equity. If humans, collectively, are doing harm to the planet and to each other, who should have to sacrifice or undertake disproportionate effort to rectify the situation? It is not even clear whether breaking questions of equity down by country is itself equitable or ethical. In a way, each country has a responsibility to live within its own geographical constraints. America consumes too much, and the Chinese cannot consume what Americans can, because China has too many people. A spatially explicit assessment of population and our planet's ecological carrying capacity drives a difficult ethical conversation. This conversation must be front and center as we discuss the future of our species and of our planet.

This will not be an easy process. All of us, to some extent, are ignorant of the range of issues at play in this difficult assessment. I certainly learned of far too many complicated and unsavory issues during the exploration that led to this book. Ignorance is a fundamental part of the human condition, yet hopefully we all learn that consistently chipping away at our own ignorance enables us to help make the world a better place. Some, to be blunt, are mentally incapable of understanding some topics. They are not ignorant. They lack certain mental faculties. In truth, this is not a material percentage of the world's population. Others do not fail to understand the issues due to ignorance. They actually understand the issues and understand the ethical issues involved. Nevertheless, they choose to ignore the facts and engage in actions that harm others while advancing their own self-interest or their own whims. Whether they are evil, merely sociopaths, or have an overdeveloped sense of self and an underdeveloped sense of empathy, differs person by person. Do not think that there are not evil people out there. There are. But sometimes they are difficult to differentiate, in their resulting behaviors, from the willfully ignorant, or even the stupid. Far too many people embrace willful ignorance when thinking about some of the big issues that face our species, our society, and our planet. Willful ignorance is a strategy that is just too easy to embrace, particularly when you feel that your self-interest might very well be put at risk if you truly understood the facts. Often, it is easier to simply go about life, while closing off your own curiosity about the larger world, forgoing the inconvenient truths you might encounter. Unfortunately, the proportion of humans in this category is larger than we might like to hope, and far too many with a will to power fall in this category. How many stray into a more toxic and recalcitrant form of self-delusion is hard to estimate, but this too is part of our reality that makes positive change more challenging than it need be.

Part of the collective ethical dialogue that we must have must recognize that not everyone in the discussion is a "good actor" who is genuinely invested in the wellbeing of our species, our society, or our planet. There are bad actors—both people and organizations—and they must be named, shamed, and neutralized. Do not confuse ethical behavior and polite behavior. They do not always intersect or reinforce each other. Some ethical goals, such as the fate of our species and our planet, are worth fighting for. And fighting can be messy.

Do not be deceived by the polite who seek their own comfort and self-interest at the expense of harm done to others and our planet. Do not immediately dismiss the impolite and abrasive who fight to reduce harm to others and our planet. Do not be fooled by the incendiary who says everything will be fine as long as "those people" just leave the "good people" alone to make the world great again. Progress is messy, and although we may all hope that the arc of history bends toward justice, in truth, this happens only if we all commit to a quest of ethical action toward each other and toward the planet that gives us life.

A Call for Participation

This book is a call for participation. The "cookbook" for global leaders and global citizens offered in Chapter 12 provided a number of ways in which everyone on this planet could get involved in a "Planet of 3 Billion" (P3B) dialogue. Although words alone have a certain power, the words in this book have been given particular meaning by the hard work that so many people have put into developing the rich spatio-temporal (i.e., geographic) data sets that underpin and illuminate the arguments in this book. Despite all of that hard work, there are still significant gaps in the data. Although I have much confidence in the conclusions of this book at the macro scale, I have the same confidence that micro- and meso-scale detail needs to coalesce in the digital geographic data resources if we are to help make the argument to various national, regional, and local publics.

If a picture is worth a thousand words, I would argue that a map can be worth several thousand. An animated map, comprising rich spatio-temporal data, that can itself be interrogated can then be worth tens of thousands. These words are not just facts about our world. They also form and provoke important questions about our world. Such animated maps, I would argue, are intellectual efficiencies that enable everyone, expert and layperson alike, to embark from a common point on an intellectual and moral journey to understand the world around us in a way that empowers practical action, both strategic and tactical in nature.

As such, please take this as a call to participate in the curation of key digital geographic data sets about how the towns and cities, regions and ecosystems, and nations that you care about have evolved over time, and how they have come together to shape the world that we live in today. The challenges that we face are local, regional, national, and global all at the same time. As I said earlier, we must think globally and act locally, but we must also act regionally and globally too. If we are to be confident in our actions, we need geographic data that can underpin a shared understanding.

This is one of the reasons that my colleagues and I created MapStory.org, the atlas of change that anyone can edit. It is a place where anyone and everyone can organize and share what they know about how our world has changed over time—whether long ago or just yesterday—at local, regional, national, and global scales. Like Wikipedia, it is a free and open platform that everybody is invited to leverage as a global platform to amplify their own voices, while inviting others to join in the quest to create comprehensive, data-driven understanding of various phenomena. Rather than collaboratively writing and editing encyclopedia articles, MapStory lets you collaboratively create and edit geographic data that demonstrate change over time, no matter whether it is geologic time, prehistory, ancient history, or the recent centuries over which modern industrial humanity has come into its own while devastating the planet.

I encourage you all, whether scientists, social scientists, students, geographic enthusiasts, concerned citizens, or policymakers to organize and share what you know about how our world has evolved, including the very significant ecological challenges that we as a species have created. I also encourage you to think of this as a collective, collaborative endeavor in which you can invite your friends and colleagues to help you build out a particular data set (what we call StoryLayers), or to build various narratives (what we call MapStories) atop the data you and others have created.

It is about time that we all apply our limitless creativity to building a collective understanding of the planet that we call home. And there is no better way to achieve understanding than contributing to a project that is near and dear to our heart. So, follow your passion, and harness all of the "cognitive surplus" and surplus of goodwill that you have and bring it to bear on how we together can understand the world around us and the grave challenges that our species and our planet presently face. The more we all know about the story of our planet, the more likely we are to treat it, and each other, well.

www.Planet3Billion.com

The New Geographic Data Needed to Navigate to the New Population Plateau

Building on the work of Wackernagel and Sutton and their respective groups of colleagues, this book has discussed the kinds of spatio-temporal observations (i.e., geographical observations over time) that would need to exist in rigorous digital form in order to calculate Earth's actual carrying capacity. Although I stand by my estimate that Earth can carry no more than 3 billion modern industrialized humans without accruing long-term ecological debt, a substantial data collection effort is required to answer all of the local-, regional-, and national-level questions that this analysis precipitates.

If we are to provide concrete guidance to citizens, policymakers, and business leaders with meaningful implementation guidance on which half of the planet (or potentially more) must be set aside for nature, the development of rigorous spatio-temporal data sets should be considered a pressing priority for the global community. Otherwise, some will be paralyzed by the incompleteness of data at local scales, despite the overwhelming global evidence that hits them between the eyes through basic "intraocular analysis." The evidence is clear, if not precise, that the human population will need to decrease in total number, and in its per capita ecological footprint, in order to unwind the ecological folly of the 20th century.

Yet even those who see that immediate action is necessary will require high-resolution data that provide overwhelming "local" evidence to help make the case to leaders and citizens in every corner of the planet. Moreover, as a practical data resource, these data will guide our strategies as we take action to move to a new, lower, population plateau. As such, it is worth discussing in some detail five global-scale data projects that I believe we must undertake before we can make such rigorous calculations.

I. High-Resolution Historical Ecoregion Data

The existing global ecoregions data set generated by RESOLVE in 2017 is invaluable. However, it should be considered just a starting point for more rigorous data collection. It was built to adhere to the same level of detail as the U.S. EPA's Level III Ecological Regions of North America data set. The next generation of this data effort should address at least three additional points.

First, as discussed earlier, this data set represents only a snapshot of the historical change that occurred over time, and the geographies defined in this data set do not capture the aggressive human-induced ecological change that has occurred over the past 200 years—not to mention the preceding 16,000 years. It fails to capture the ways in which human slash-and-burn practices and subsistence farming have eliminated massive tracts of historical ecosystems, fundamentally changing their boundaries. It also fails to capture the profound transmogrification of hydrological resources, such as the Lake Chad Basin.

Second, such a data set must also address the oceans and the various transitional marine ecosystems that represent critical breeding grounds, ecological filters, and freshwater recharge zones. The vast majorities of our planet's surface and its volume are oceans. Without understanding these ocean ecosystems and how they connect to and serve our terrestrial ecosystems, as well as our atmosphere, we are nowhere. The nascent maritime ecoregions work cited in this book requires significantly more investment, including investment in better characterizing the relationships with adjacent terrestrial ecoregions.

Third, both the maritime and the terrestrial ecoregions data sets require further investment in order to bring our understanding of ecoregions, worldwide, up to par with the EPA's Level IV ecoregions data standard, which offers an additional level of granularity in the description of ecoregions. This Level IV data set provides superior insights into how ecological resources can be managed, and there is no global ecoregions data set yet available at this level of granularity. Yet, even the EPA's Level IV work (like the global ecoregions work by WWF and RESOLVE) is done at a geographic scale of 1:3,000,000, which elides over critical ecological details. Though a herculean task, this data set would be built at a much finer geographic scale, say 1:25,000—which laypeople can understand as providing detailed information at a human scale within their own communities. At this scale, we will find all manner of essential ecological subsystems with distinct geographies that need protection if we expect them to return to us the ecosystem goods and services that we have depended on historically. Anyone who claims to have an interest in understanding, managing, and protecting our planet's wellbeing should be a vocal proponent for this kind of rigorous, spatio-temporal data initiative.

II. Mapping Industrialization and Waste

There is a considerable range of geographical data sources related to the industrialization of the global landscape and the accumulation of persistent wastes across it. Unfortunately, each one of these data layers must be pieced together from a host of poorly organized data sources. Yet if we are to understand how humanity's footprint has compromised Earth's ecoregions, there is a need to rigorously map at a 1:25,000 scale each of the five types of industrialization undergone and the five types of persistent waste accumulated over the past 200 years, as detailed in Chapters 4 and 5, including the flows of natural resource consumption demanded by each industrial process. Each of the five types of industrialization poses different challenges to anyone who seeks to map their geographic evolution. Important parts of the history of extraction and combustion have been mapped at different levels of detail across the historical record. For instance, the geographical history of deforestation, agricultural development, and urbanization has been recorded and reconstructed going back some 16,000 years.

Obviously, the mapping of these processes has become increasingly precise over the past four decades, with the rise of satellite remote sensing and its improving ability to collect higher-resolution data. The rise of manufacturing and modern infrastructure has been recorded rather comprehensively over the past few hundred years, though this historical record is often quite fragmented, requiring significant effort to reconstruct. Waves of industrial development have seen older facilities and infrastructure lost to the sands of time, thus requiring historical reconstruction on the basis of land use records, ownership records, and taxation records. Some industries demonstrate solid institutional memories. Others do not.

The geographical evolution of persistent human wastes is an altogether different problem. The scientific sampling of many of these wastes is a relatively recent endeavor that has been undertaken unevenly, both geographically and temporally, as new public bureaucracies have emerged to take on this job. Yet in a way, it is easier to reconstruct the historical spread of these wastes, as humanity's invention of many of these persistent pollutants is historically recent, in events that are well recorded. Still others, such as the five ocean garbage patches, eluded human detection until the 1990s, after which we realized that they had already grown to continental proportions. As such, considerable effort must be applied in order to assemble quality geographical data sets of industrialization and of the accumulation of persistent wastes, if we are to properly characterize the history of humanity's ecological footprint.

III. Historical Depletion of Ecosystem Goods and Services

It is critical that we estimate the variety and volume of ecosystem goods and services presently and historically generated by a given ecoregion, the relational flows between ecoregions, and how they have changed as ecoregions have been forced to support the ecological burden of industrialization, waste, and human predation. This estimate would, obviously, be based on the high-resolution ecoregions mapping effort just described. This is no simple task, as it requires first mapping the present-day production of ecosystem goods and services—a task that many academicians and public servants have aspired to for some time, with only limited success. Then, after the patterns of industrialization and accumulation of persistent wastes are mapped, historical reconstruction of these resources would need to be completed. In many geographies, there are well-documented depletions of ecosystem goods and services, whether due to the extinguishment of particular fauna, the denuding of particular flora, the diversion of water resources, or the destruction of the local atmosphere or microclimate. To truly calculate the carrying capacity of the planet, it will be necessary to understand the historical baseline production of ecosystem goods and services so as to properly characterize the ecological debt that has accumulated over the past couple centuries. Only then can we determine what efforts humanity might take to replenish and make resilient previously compromised ecological resources and how many additional humans that might allow the planet to support.

IV. Growth in Animal Populations

Any effort to map ecosystem goods and services would certainly count wild animal populations and the ecological processes driving these populations to wax or wane. However, the domestication of animals for food and companionship has grown explosively, tracking the growth of the modern human population. If anything, the growth in living standards has actually made the domesticated animal population grow at a rate faster than that of the human population. And the ecological implications have been profound. Although many ecoregions long managed to sustain large communities of natural wildlife, modern animal domestication has demanded the concentration of large animal populations whose consumption and wastes have been riven from the ecosystems from which their wild progenitors arose. If we are to understand the ecological impacts of the more than 20 billion domesticated animals that humans have chosen to raise on this planet, the growth of these populations must be mapped over time. Which ecoregions have most been damaged by animal domestication and the geographic patterns of animal consumption must be understood if we are to formulate any meaningful strategies for feeding and entertaining humanity while rebuilding ecological capacity.

V. Estimated Human Footprint by Ecoregion over Time

The art of mapping human population density on a global basis has gone through many refinements over the past couple decades. However, the ecological impacts of human populations differ over different geographies. Just because population density (per square kilometer) is the same in two places does not mean that the ecological footprint of those humans is the same in the two geographies. Whether the landscape has been engineered for industrialization or to deal sustainably with wastes can lead to completely different ecological footprints for the same population densities. This effort will shine a light on where population growth has outstripped the sustainable engineering foundation needed to lighten the human footprint. In mapping this footprint, the challenge will be to include estimates of how the populations of each region drive the ecological footprint through consumption from global supply chains, so that the ecological pressures on distant geographies can be tied to the point of consumption demand.

Letters to the Pope, Jeff Bezos, Bill and Melinda Gates, and President Xi Jinping

Holy See,

To the casual observer, it has appeared that the Catholic Church has long been an advocate for filling Earth with more and more people. Perhaps it is more accurate to say that the Church has simply steadfastly defended its doctrine prohibiting the use of various family planning technologies, regardless of the implications for the population trends this policy might induce. However, since the middle of the 20th century, as the world's population has careened out of control, each of your predecessors has defended Church doctrine, telling policymakers that any population control policy must adhere to Catholic doctrine or they will suffer the political consequences.

In your May 2015 encyclical, Laudato Si, you followed their lead, indicating that "demographic growth is fully compatible with an integral and shared development." You continued, "To blame population growth instead of extreme and selective consumerism on the part of some, is one way of refusing to face the issues." While it is no doubt true that modern consumerism has enormous negative ecological implications, I would like to engage you in a dialogue about your assumptions on population.

In Laudato Si, you called for all people of the world to take "swift and unified global action" against environmental degradation and global warming. As a world leader with a massive global following, you, indeed, took a bold step in the right direction. Thank you for that. Your accompanying statements on population, however, seemed to ignore the assertions of your own Pontifical Academies of Sciences and Social Sciences, released one month prior to your encyclical. The Pontifical Academies were clear that there is a knowable sustainable global population plateau that needs to be met and maintained. Although they were silent on the actual number of this population plateau, the achievement and maintenance of such a plateau will necessitate the Church's reconsideration of its stance on family planning technologies.

In this context, I ask you, with all due respect, what do you believe that new population plateau to be? Or do you believe that world population could grow indefinitely and still not suffer the ecological consequences of which you have warned the world? I suspect not. My assessment holds that our planet's actual carrying capacity can support only roughly 3 billion humans, with modestly modern lifestyles, if we are to establish long-term planetary resilience. I have made the geographic data that I marshaled in this analysis of our finite planet available to you and to your Pontifical Academies. I encourage you to engage these and other data to independently develop your own assessment of the population plateau that you believe would make possible a resilient planet that can sustain humanity over the long term. I look forward to discussing any divergence in our assessments.

Your passion for the wellbeing of our planet and its people is clear. I applaud your willingness to speak truth to power, as it relates to ecological destruction, which has been on overdrive for the past century. Too many of our corporations and governments alike appear to favor profit and hegemony over the long-term wellbeing, prosperity, peace, and security of humanity. They are singularly focused on growth. Might I suggest that the Church's doctrinal support of this growth mindset, though perhaps inadvertent, runs counter to your calls for ecological resilience?

I ask you to consider my arguments and their supporting data with an open mind, and to think both historically and geographically as you develop your assessment. Then, I beseech you, make your views known to the world. If doctrinal reconsiderations are due in order to bring our global population to a new, lower plateau so as to save our Earth from ecological destruction, then please, show the same leadership that you have already to help us get there as quickly as possible.

Thank you in advance for your consideration.

Dr. Christopher K. Tucker

Dear Mr. Bezos,

In your 2017 speech at the National Space Club's Goddard Memorial Dinner in Washington, DC, you invoked the need for humanity to become a space-faring species if it were to have the physical room necessary to fulfill its potential. As you accepted the Dr. Robert H. Goddard Memorial Trophy on behalf of Blue Origin's New Shepard team, you said that only space would give humanity the room to achieve the size necessary to have "a thousand Einsteins." In the audience, I applauded your speech.

After your talk, I was left to wonder what your assumptions were regarding our global population challenges and the dire ecological consequences that they have for our planet. Although it might someday be amazing to have a human population in the trillions, scattered across a seemingly infinite universe, we currently have a single planet playing host to some 7.5 billion people, on a trend line to soon hold 9 or 11 or even 13 billion people. Population is certainly growing faster than humans can currently be launched into space, and as of yet there are no permanent galactic destinations for them to reach. So, although envisioning a distant future of a universe with a human population of near infinite size is inspiring, it also appears to distract us from the dire consequences of our immediate population predicament.

Obviously, it is not your job to solve the world's population challenges. However, you are responsible for one of the largest commercial enterprises ever to appear the face of Earth, which is quickly becoming a proximate cause for much of the deluge of postconsumer waste in our daily lives. In addition, given the scale of your business, how you design this enterprise has epic ecological consequences for our planet. The choices that Amazon makes in the supply chains it creates will have profound impacts on the global landscape, the resilience of our planet, and its ability to sustain human life.

As the CEO of Amazon, you no doubt have sized the global market for your goods and services, and have a sense of its natural limits. I would ask you to take the same keen eye and try your hand at calculating the carrying capacity of Earth, to determine how many people you believe it could hold without incurring long-term ecological debt. If you, like I, see that humanity has overshot our planet's ability to sustain us without incurring such debt, I would ask you what your guidance would be to business leaders if indeed we were successful at bringing our population down to a lower, more sustainable level, as they will be forced to navigate a world of de-growth.

Thank you in advance for your consideration.

Dr. Christopher K. Tucker

Dear Mr. and Mrs. Gates,

You wield perhaps the largest philanthropic tool ever to grace human civilization. Your Foundation's focus on eradicating global epidemics and endemic diseases has transformed how we think about wellbeing and prosperity across the developing world. Your foundation's investments have also raised awareness about and provided access to family planning technologies, demonstrating your recognition that they are essential to women's empowerment and to building strong families and communities. The result of this admirable, godly work will be more souls on our planet, unless these same developing nations embrace new fertility norms.

As I see it, your investment in women's empowerment also has the potential to pay dividends in helping blunt the coming population explosion in Sub-Saharan Africa. As you likely already know, according to the United Nations Department of Economic and Social Affairs' updated medium-fertility scenario released in 2013, the global population will rise from just over 7 billion in 2012 to nearly 9.6 billion by 2050, and roughly half of this growth will occur in Sub-Saharan Africa alone. Many miss the fact that this population growth also promises a level of ecological destruction in some of the world's most critical ecoregions, which generate ecosystem goods and services that are critical to our species and the planet that sustains us.

My analysis, and that of other scholars in the field, holds that our human population has already exceeded the carrying capacity of our planet. Regardless of the specific calculation, if you find that at 7.5 billion we already have exceeded our planet's ecological boundaries, it is hard not to focus on the ecological debt that we have been accruing, which will come due with disastrous consequences if measures are not taken to bring human population in balance with our planet's ecological capacity.

Given your thoughtful philanthropic strategy on so many issues, I ask whether you have undertaken your own analysis of the world's carrying capacity? If so, what is your number? And if it is below 7.5 billion, how do you see us collectively addressing these issues in a thoughtful, strategic, and humane manner?

Thank you in advance for your consideration.

Dr. Christopher K. Tucker

President Xi,

Some say that your greatest obstacle to completing China's rise to global wealth and power may be a greying population. Some suggest that you may soon bring a historic end to China's roughly four-decade-old policy of limiting the number of children each family can have, possibly even creating new incentives for families to have more children. Though it is no doubt good for China to evolve its approach to its runaway population growth, the notion that China and Planet Earth are best served by incentivizing additional population growth is deeply misguided. Our planet's long-term ecological carrying capacity can support only 3 billion humans, by my estimate, and the ecological debt that humanity's improvident population growth has incurred is coming due in dangerous ways.

In this context, might I suggest that the metrics by which China currently assesses its future success are ill advised? Might I suggest that China is being misled by economists who are wrongly obsessed with continuous GDP growth and the relative size of China's economy compared with that of the United States? Might I suggest that the next steps in China's modernization must navigate a future that simply was not contemplated when you began your national journey toward modernization? The ecological consequences of China driving further modernization through continued population growth will precipitate untold harms to China and to the entire planet.

Simply put, a Chinese population of nearly 1.5 billion, with nearly half a billion in the middle class or more affluent classes, is capable of annihilating virtually any species or ecosystem on Earth, in very short order, with a simple muscle movement. Due to its sheer numerical size, a minor shift in Chinese consumer patterns can have devastating effects on the planet.

Almost single-handedly, China has managed to crush shark populations across our oceans, killing nearly 100,000,000 sharks a year for their fins in order to provide shark-fin soup to its growing population. Growing Chinese affluence has also managed to crush African rhinoceros populations, to feed the demand for rhinoceros horn inspired by traditional Chinese medicine. At home in the South China Sea, Chinese overfishing has destroyed vital ocean ecosystems, reefs have been devastated in order to sell coral art and trinkets, and the poaching of giant clams has undermined critical ocean ecosystems so that they can be sold in China as aphrodisiacs, jewelry, and home decorations. I could go on and on. But the long list of species and ecosystems that are being annihilated by the consumption of the large and growing Chinese population is not the point. And although global climate change is set to eliminate much of the coastal urban wealth that China has accumulated over the past half-century, that also is not the point.

The point is that the "empty-world vision" animating your economic advisors, and those of the world's other political leaders, fails to recognize that our global population has already begun to undermine the long-term viability of the ecological resources that sustain our own species' life on this planet. Our economic growth has become "uneconomic growth." And increasing our global population will only magnify this dynamic.

Your economic advisers would have us not account for humanity's deletion of vast amounts of natural capital, nor for the massive accumulation of persistent wastes as part of the marginal disutility of our economic growth. Your advisers would have us focus on the economic value added to our finite planet by human labor and capital, without paying attention to the natural processes that these investments have come to devastate, as our global population has outstripped our planet's ecological carrying capacity.

China, it seems, is the single most important player in defining the future of our species and the planet that sustains us. China can help lead humanity to a new, lower population plateau, while also increasing the individual productivity, wealth, and wellbeing of Chinese citizens, all while navigating the resulting era of economic de-growth. China can help humanity innovate its way to a more sustainable future that allows our planet to rebound from the ecological devastation that has come from the past two centuries of industrialization.

Or, China can grow its population further and become, perhaps, the single largest force in sending our planet over the ecological brink. It may be that you and your advisers disagree with this assessment. If so, I am curious—what are their calculations of Earth's carrying capacity? How many people do you believe Earth can support? If not, I look forward to China's leadership in helping our species navigate our way to a global population of 3 billion, with dignity, prosperity, and security.

Thank you in advance for your consideration.

Dr. Christopher K. Tucker

Acknowledgments

During the process of writing this book, I have benefited from the tolerance and insights of many. The tolerance of my wife and family, as I impulsively added this project to my already chronically overcommitted schedule, was over and above the call, always with the kind of love and support none of us have the right to expect. Beyond the good fortune of enjoying an amazing support system at home, I must convey the contributions of all those who helped me realize the best parts of this book.

Any heights of insight that this book has achieved are due only to the intellectual giants on whose shoulders I was able to climb. Any and all shortfalls and errors are due to my own sloth, ignorance, and oversight.

This book began with a midnight email exchange with my friend and colleague Parag Khanna, as he was gallivanting across some distant part of the globe, as is his wont. After reading about the status of his latest book project, I mentioned that my various riffs and rants had congealed into a rough book concept that I had a great passion to pursue. He graciously encouraged me to tell him all about it. The resulting very long email, filled with run-on sentences and my exhortations about why particular chapters would need to be written in particular ways, led Parag to give me marching orders to get started immediately. That long email, with few exceptions, served as the outline of the book you see here before you. Parag has kindly answered endless trivia questions from this first-time author, and I could not have navigated the publishing process without his sage advice. To Parag, I pass along my thanks and my hopes for future collaboration.

I owe my brother, Jonathan Tucker, a debt of gratitude. First, it is he to whom I owe my early academic interests in science and technology policy—that is, the policies that shape the rate and direction of scientific advance and technological change. For sure, this was a peculiar interest and ambition for an 18 year old entering college. But it has shaped everything I have done academically and professionally over the subsequent (nearly) three decades. It was Jon, whether he knows it or not, who was the spark that led me to apply to Columbia University for college, where my early years under the tutelage of Richard Nelson and Michael Crow fundamentally shaped the worldview and arguments in this book. It was also Jon who read the very first incomplete and very imperfect draft of my manuscript, and encouraged me to continue on with the same brash voice with which I began. To Jon, I thank you for the gifts you have given me over the decades, not the least of which is simply being a great big brother.

Soon after I launched into this project in earnest, I confided in Jonathan Marino, who has been a stalwart partner in crime over the past five years. Without Jon, my vision for MapStory.org would never have succeeded. After I kicked off MapStory with my friend and colleague Robert Tomes, Jon asked to come aboard our little social venture and has helped it become a place on the World Wide Web where all people can organize and share their knowledge about our world in space and in time. Jon, wise beyond his years, has been a confidante on many of my schemes, and provided priceless encouragement and support as I began devising what I saw as much more than just a book. The curation of P3B's digital data sets and their presentation as a larger architecture of participation owes much to Jon. We hope that this platform will help enable global debate about and exploration of the issues raised in this book for generations to come. To Jon Marino, thanks for always being there when I have a new urge to instigate.

Serendipity is always convenient. And when it strikes in a hot tub in Mexico while drinking beer with friends, even better. As I hit my stride in writing, I was lucky to find myself talking with my friend and colleague Ben Tuttle, who I knew as an outstanding remote-sensing geographer, technology innovator, national security professional, map tattoo aficionado, and all around good guy. It just so happened that he also did his doctoral work under Paul Sutton at the University of Denver, whose work plays prominently in this book. In chatting about my book, Ben at first seemed overly engaged for vacation banter. After I explained the remote-sensing literature that I thought would be core to my arguments in Chapter 8: How to Calculate Earth's Maximum Carrying Capacity, he laughed and began texting me links to papers he had co-authored with Paul Sutton, and for which he had done much of the data work in his previous life as a graduate student. Who would have guessed? He then was kind enough to make introductions to Professor Sutton, a connection which ultimately paid great dividends for this book. Paul was nothing if not gracious with his time and cheered me on, asking when the book would be ready for him to assign to his students. Beyond this encouragement, Paul's willingness to go back and forth with me on this crucial chapter provided me with the confidence needed on the home stretch.

I was fortunate to have come across Mathis Wackernagel and his work with the Global Footprint Network when my friend Barbara Ryan, as the Executive Director of the Group on Earth Observations, had me come and speak at the 2014 Eye on Earth symposium in Abu Dhabi. His talk acquainted me with this powerful approach to calculating ecological footprint at a national level. Subsequently, he was generous with his time and sympathetic toward my project. His bluntness and transparency about his methodology and how it relates to others, such as Sutton's, helped me solidify my arguments in important ways. I only hope that I fairly represented his important work, and that my friendly amendments are taken positively.

Thanks to my research assistant, Hettie "Sudie" Caitlin Brown—and thanks to her professor and American Geographical Society President Marie Daly Price for recommending her to me. Sudie's tireless bird-dogging of critical sources helped me ensure that my arguments were up to date with the frontiers of scientific knowledge. It was Sudie who brought to me the latest ecoregions work by RESOLVE and the scholarly articles situating this crucial data set within the latest conservation debates. This rich and powerful literature helped anchor my arguments but raised many questions. Eric Dinerstein, Director of Biodiversity and Wildlife Solutions at RESOLVE, and former Chief Scientist and Vice President for Conservation Science at the WWF, took the time to meet, answer my outstanding questions, and point me to even more data. The more I spend time with this body of knowledge, the more I am inspired by the work of this community of biogeographers and impressed by the criticality of this scholarship to the future of our planet.

I would be remiss if I failed to mention my colleagues at the American Geographical Society and all of the inspiring speakers who have been a part of our Geography2050 symposia—a multiyear strategic dialogue about the vital trends that will reshape the geography of our planet over the coming decades. It is this cauldron of ideas that first spurred me to work out the central thesis of this book. Leading thinkers in demography, urbanization, mobility, climate change, sustainability, conservation, energy, and so many other crucial topics helped me tackle the complexities inherent in the arguments underlying this book. Thanks to all our AGS Councilors for making that experience possible. Thanks to AGS President Emeritus Jerry Dobson for recruiting me for the AGS Council, and to AGS Chairman Emeritus John Gould for seeing in me the potential to follow in his footsteps. Thanks to AGS President Marie Daly Price for her partnership. And thanks to AGS CEO John Konarski for making Geography2050 and all that we do at the Society successful.

Without Rey Dizon, this book would just be 78,000 words in a GoogleDoc. With Rey, it has become so much more. After having had the good fortune of meeting Rey years ago as a MapStory Summer Fellow, I was even more fortunate when he graduated to be Creative Director for the MapStory Foundation. This has given me opportunities to collaborate with Rey on his countless feats of cartographic wonder. I was lucky that Rey was ready, willing, and able to help with this book project. As it turns out, he is as talented in the dark arts of online digital cartography as in the creation of print cartography and publishing. These are very different talents, both of which are critical to this project. Thanks to Rey for making time in your busy schedule to help convey my arguments to a wider public in a visual and compelling way.

Even with great design and cartography, there is a thin line between a manuscript with potential and a book that meets the high standards of your reading audience. Thanks to my copy editor, Lise Lingo for getting me over that line. And, to Eleanor O'Connor for steering my team in the right direction. Of course, thanks to Barb Faculjak and Liz Lyon (@geographerliz) for connecting me with the amazing women in their amazing networks.

Thanks to James Scrivener for helping ensure that my energy calculations were solid. Everyone writing about energy trends should have a solar CEO who is savant-like in his grasp of energy statistics. Any first-time book writer should also be so fortunate as to have a neighbor who is an accomplished author, to provide that key conversation over coffee and occasional gentle inquiries as to your progress while walking kids to school. My neighbor Claudia Kalb, author of *Andy Warhol Was a Hoarder: Inside the Minds of History's Great Personalities*, is not only a delight to be around, she also gives great advice. And when your book is driven by geography, you could only be so lucky as to have Alec Murphy give your manuscript a read. He is always a source of good counsel and inspiration. But when your book is meant to appeal to readers of all ages who are interested in the world around them, concerned about the fate of the planet, and who want solutions, not just problems—it has been invaluable to have so many friends from so many walks of life give this manuscript a read and provide blunt feedback. Thanks to all.

My penultimate thanks is to my agent, Rafe Sagalyn at ICMPartners, for seeing in this project the same potential as did I. His willingness to take me on gave me the confidence I needed to bring this project to completion.

I must end as I began. This book is just one more thing that I never would have been able to accomplish without the love and support of my wife and family. My parents, who live nearby, are a continued source of aid, whatever I need, whenever I need it. My children, Lily and Ava, are a source of inspiration, as they tackle the world we have left for them. And my wife and love of my life, Ann, has made me a much better man than I would otherwise be.

To all, I am indebted. And to the extent that you found value in this book, it is in no small part because of their contribution.

Bibliography

Albeck-Ripka, Livia. 2017. Are We an Invasive Species? Plus, Other Burning Climate Questions. *New York Times*, December 6.

American Museum of Natural History. 2016. Human Population Through Time (video). https://www.youtube.com/watch?v=PUwmA3Q0_OE. November 4.

American Museum of Natural History. 2018. Seven Million Years of Human Evolution (video). https://www.youtube.com/watch?v=DZv8VyIQ7YU. November 3.

Bailey, R. G. 1976. *Ecoregions of the United States* (map). 1:7,500,000 scale. U.S. Department of Agriculture, Forest Service, Intermountain Region, Ogden, Utah.

Bailey, R. G. 1980. Description of the Ecoregions of the United States. United States Department of Agriculture, Forest Service, Washington, DC. Miscellaneous Publication Number 1391.

Bailey, R. G. 1983. Delineation of Ecosystem Regions. *Environmental Management* 7:365-373.

Bailey, R. G. 1995. *Ecosystem Geography*. New York: Springer-Verlag.

Bailey, R. G. 2002. *Ecoregion-Based Design for Sustainability*. New York: Springer.

Bailey, R. G. 2009. *Ecosystem Geography: From Ecoregions to Sites*, 2nd ed. New York: Springer.

Bailey, R. G. 2014. *Ecoregions: The Ecosystem Geography of the Oceans and Continents*, 2nd ed. New York: Springer.

Bailey, R. G., P. E. Avers, T. King, and W. H. McNab. 1994. *Ecoregions and Subregions of the United States* (map and table). 1:7,500,000 scale. U.S. Department of Agriculture, Forest Service, Washington, DC.

Bailey, R. G., S. C. Zoltai, and E. B. Wiken. 1985. Ecoregionalization in Canada and the United States. *Geoforum* 16(3):265-275.

Biello, David. 2008. Oceanic Dead Zones Continue to Spread. *Scientific American*, August 15.

Biodiversity Information System for Europe. Mapping and Assessment of Ecosystems and Their Services (MAES). https://biodiversity.europa.eu/maes.

Bittman, Mark. 2008. *Food Matters: A Guide to Conscious Eating*. New York: Simon and Schuster.

Botequilha Leitão A, J. Ahern, and K. McGarigal. 2006. *Measuring Landscapes. A Planner's Handbook*, 2nd ed. Washington, DC: Island Press.

Boulding, Kenneth E. 1966. The Economics of the Coming Spaceship Earth. In *Environmental Quality in a Growing Economy*, ed. Henry Jarrett. Baltimore: Johns Hopkins Press.

Box, Jason E., William T. Colgan, Bert Wouters, David O. Burgess, Shad O'Neel, Laura I. Thomson, et al. 2018. Global Sea-Level Contribution from Arctic Land Ice: 1971-2017. *Environmental Research Letters* 13(12):125012.

Brewer, Peter, and James Barry. 2008. Rising Acidity in the Ocean: The *Other* CO_2 Problem. *Scientific American*, September 1.

Brinkley, Douglas. 2010. *The Wilderness Warrior: Theodore Roosevelt and the Crusade for America*. New York: Harper Perennial.

Brown, Lester R., Gary Gardner, and Brian Halweil. 1998. Beyond Malthus: Sixteen Dimensions of the Population Problem. Worldwatch Paper 143, Worldwatch Institute, Washington, DC.

Bryant, Lee. 2015. Ocean 'Dead Zones' Are Spreading – and That Spells Disaster for Fish. *The Conversation*. April 7.

Bryce, S. A., James M. Omernik, and D. P. Larsen. 1999. Ecoregions – a Geographic Framework to Guide Risk Characterization and Ecosystem Management. *Environmental Practice* 1(3):141-155.

Burkhard, Benjamin, and Joachim Maes, eds. 2017. *Mapping Ecosystem Services*. Sofia, Bulgaria: Pensoft Publishing.

Carson, Rachel. 1962. *Silent Spring*. Boston: Houghton Mifflin.

Cohen, Joel E. 1995. Population Growth and Earth's Human Carrying

Capacity. *Science* 269(5222):341-346.

Cohen, Joel E. 1996. *How Many People Can The Earth Support*. New York: W. W. Norton and Company, Inc.

Crossman, N. D., Benjamin Burkhard, S. Nedkov, Louise Willemen, K. Petz, I. Palomo, et al. 2013. A Blueprint for Mapping and Modelling. *Ecosystem Services* 4:4-14.

Cumming, Vivien. 2016. How Many People Can Our Planet Really Support: We Do Not Know If Today's Population of Seven Billion Is Remotely Sustainable, or What the Limit Is. BBC Online, March 14.

Dailey, Gretchen C., Anne H. Ehrlich, and Paul R. Ehrlich. 1994. Optimum Human Population Size. *Population and Environment* 15(6, July):469-475, doi: 10.1007/BF02211719.

Daly, Herman E., and Joshua Farley. 2004. *Ecological Economics: Principles and Applications*. Washington, DC: Island Press.

Darwin, Charles. 1859. *On the Origin of Species*. London: John Murray.

Davis, Mike. 2005. *Planet of Slums*. London: Verso Books.

De Araujo Barbosa C. C., P. M. Atkinson, and J. A. Dearing. 2015. Remote Sensing of Ecosystem Services: A Synthetic Review. *Ecological Indicators* 52:430-443.

De Menocal, P. B., and J. E. Tierney. 2012. Green Sahara: African Humid Periods Paced by Earth's Orbital Changes. *Nature Education Knowledge* 3(10):12.

Diamond, Jared. 2005. *Collapse: How Societies Choose to Fail or Succeed.* New York: Viking Press.

Dinerstein, Eric, David Olson, Anup Joshi, Carly Vynne, Neil D. Burgess, Eric Wikramanayake, et al. 2017. An Ecoregion-Based Approach to Protecting Half the Terrestrial Realm. *BioScience* 67(6, June 1):534-545.

Drache, Hiram. The Impact of John Deere's Plow. Illinois Historic Preservation Agency, Springfield, IL.

Dutkiewicz, S., J. J. Morris, M. J. Follows, J. Scott, O. Levitan, S. T. Dyhrman, and I. Berman-Frank. 2015. Impact of Ocean Acidification on the Structure of Future Phytoplankton Communities. *Nature Climate Change* 5(11):1002-1006, doi: 10.1038/nclimate2722.

Ehrlich, Paul R., and Anne H. Ehrlich. 1990. *The Population Explosion.* New York: Simon and Schuster.

Ehrlich, Paul R., and Anne H. Ehrlich. 1991. *Healing the Planet.* New York: Addison Wesley.

Ehrlich, Paul R., and James P. Holdren. 1971. Impact of Population Growth. *Science* 171:1212-1217.

Ehrlich, Paul R., G. C. Daily, and L. H. Goulder. 1992. Population Growth, Economic Growth, and Market Economies. *Contention* 2:17-35.

Ehrlich, Paul. 1968. *The Population Bomb.* New York: Sierra Club/Ballantine Books.

Ellis, Erle C., and Navin Ramankutty. 2008. Putting People in the Map:

Anthropogenic Biomes of the World. *Frontiers in Ecology and Environment* 6(8):439-447.

Ellis, Erle C., Jed O. Kaplan, Dorian Q. Fuller, Steve Vavrus, Kees Klein Goldewijk, and Peter H. Verburgf. 2013. Used Planet: A Global History. *Proceedings of the National Academy of Sciences* 110(20, May 14):7978-7985.

Elvidge, Christopher D., Benjamin T. Tuttle, Paul S. Sutton, Kimberly E. Baugh, Ara T. Howard, Cristina Milesi, et al. 2007. Global Distribution and Density of Constructed Impervious Surfaces. *Sensors (Basel)* 7(9, September):1962-1979, doi: 10.3390/s7091962.

Elvidge, Christopher D., Mikhail Zhizhin, Kimberly Baugh, Feng-Chi Hsu, and Tilottama Ghosh. 2016. Methods for Global Survey of Natural Gas Flaring from Visible Infrared Imaging Radiometer Suite Data. *Energies* 9(1):14, doi: 10.3390/en9010014.

Elvin, Mark. 2006. *Retreat of the Elephant: An Environmental History of China.* New Haven: Yale University Press.

Emerson, Ralph Waldo. 1836. *Nature.* Boston: James Munroe and Company.

Etminan, M., G. Myhre, E. J. Highwood, and K. P. Shine. 2016. Radiative Forcing of Carbon Dioxide, Methane, and Nitrous Oxide: A Significant Revision of the Methane Radiative Forcing. *Geophysical Research Letters* 43(24):12614-12623, doi: 10.1002/2016GL071930.

See also https://phys.org/news/2017-01-effect-methane-climate-greater-thought.html#jCp.

Farina, Almo. 2013. *Soundscape Ecology Principles, Patterns, Methods, and Applications.* New York: Springer.

Food and Agriculture Organization of the United Nations. 2006. *Livestock's Long Shadow: Environmental Issues and Options.* Rome: FAO.

Gaston, Bibi. 2016. *Gifford Pinchot and the First Foresters: The Untold Story of the Brave Men and Women Who Launched the American Conservation Movement.* New Milford, CT: Baked Apple Club Productions.

Georgescu-Roegen, Nicholas. 1971. *The Entropy Law and the Economic Process.* Cambridge, MA: Harvard University Press.

Georgescu-Roegen, Nicholas. 1977. The Steady State and Ecological Salvation: A Thermodynamic Analysis. *Bioscience* 27(4, April):266-270.

Global Footprint Network. Footprint Calculator. www.footprintcalculator.org.

Global Forest Watch. http://www.globalforestwatch.org.

Guarino, Ben. 2018. 'Hyperalarming' Study Shows Massive Insect Loss. *Washington Post,* October 15.

Haeckel, Ernst. 1866. *Generelle Morphologie der Organismen. Allgemeine Grundzüge der organischen Formen-Wissenschaft, mechanische Begründet durch die von Charles Darwin reformirte Descendenz-Theorie.*

Hansen, James. 2010. *Storms of My Grandchildren: The Truth About the Coming Climate Catastrophe and Our Last Chance to Save Humanity.* New York: Bloomsbury USA.

Hansen, James, D. Johnson, A. Lacis, S. Lebedeff, P. Lee, D. Rind, and G. Russell 1981. Climate Impact of Increasing Atmospheric Carbon Dioxide. *Science* 213(4511):957-966, doi: 10.1126/science.213.4511.957.

Hardin, G. 1968. The Tragedy of the Commons. *Science* 162(3859):1243-1248, doi: 10.1126/science.162.3859.1243.

Harris, M., and E. B. Ross. 1987. *Death, Sex, and Fertility: Population Regulation in Preindustrial and Developing Societies.* New York: Columbia University Press.

Hawken, Paul. 2017. *Drawdown: The Most Comprehensive Plan Ever Proposed to Reverse Global Warming.* New York: Penguin Books.

Heinberg, Richard. 2018. We Are Exceeding Earth's Carrying Capacity. Denying It Is Suicidal. *Quartz,* August 3.

Holdren, John P. 1991. Population and the Energy Problem. *Population and Environment* 12:231, doi: 10.1007/BF01357916.

Holdren, John P., and Paul R. Ehrlich. 1974. Human Population and the Global Environment. *American Scientist* 62:282–292.

Hotelling, Harold. 1931. The Economics of Exhaustible Resources. *Journal of Political Economy* 39(2, April):137-175.

Hwang, Andrew D. 2018. 7.5 Billion and Counting: How Many Humans Can the Earth Support? *The Conversation,* July 9.

Imeson, Anton. 2011. *Desertification, Land Degradation and Sustainability.* Hoboken, NJ: John Wiley and Sons.

Independent Evaluation Group. 2007. *The Nexus Between Infrastructure and Environment*. Washington, DC: World Bank.

Jacobs, Frank. 2019. The World's Watersheds, Mapped in Gorgeous Detail. *BigThink*, February 3.

Jiang, Li-Qing, Richard A. Feely, Brendan R. Carter, Dana J. Greeley, Dwight K. Gledhill, and Krisa M. Arzayus. 2015. Climatological Distribution of Aragonite Saturation State in the Global Oceans. *Global Biogeochemical Cycles* 29(10):1656-1673, doi: <u>10.1002/2015GB005198</u>.

Johnson, Steven. 2007. *The Ghost Map: The Story of London's Most Terrifying Epidemic—and How It Changed Science, Cities, and the Modern World*. New York: Riverhead Books.

Kaplan, Jed O., and Kristen M. Krumhardt. 2011. The KK10 Anthropogenic Land Cover Change Scenario for the Preindustrial Holocene, link to data in NetCDF format. *PANGAEA*, doi:10.1594/PANGAEA.871369, *Supplement to:* Kaplan, Jed O., Kristen M. Krumhardt, Erle C. Ellis, William F. Ruddiman, Carsten Lemmen, and Kees Klein Goldewijk. 2011. Holocene Carbon Emissions as a Result of Anthropogenic Land Cover Change. *The Holocene* 21(5), 775-791, doi: <u>10.1177/0959683610386983</u>.

Kaplan, Jed O., Kristen M. Krumhardt, and Niklaus E. Zimmermann. 2009. The Prehistoric and Preindustrial Deforestation of Europe. *Quaternary Science Reviews* 28(27-28):3016-3034, doi: <u>10.1016/j.quascirev.2009.09.028</u>.

Kareiva P., H. Tallis, T. H. Ricketts, G. C. Daily, and S. Polasky. 2011. *Natural Capital: Theory and Practice of Mapping Ecosystem Services*. Oxford, U.K.: Oxford University Press.

Kedrosky, Paul. 2011. What Scientific Concept Would Improve Everybody's Cognitive Toolkit? Shifting Baseline Syndrome. *The Edge*. https://www.edge.org/response-detail/10755.

Kendall, Edward C. 1959. *John Deere's Steel Plow*. Washington, DC: Smithsonian Institution.

Khanna, Parag. 2016. *CONNECTOGRAPHY: Mapping the Future of Global Civilization*. New York: Random House.

Kharas, Homi, and Kristofer Hamel. 2018. A Global Tipping Point: Half the World is Now Middle Class or Wealthier. *Future Development*, Brookings Institution, Washington, DC, September 27.

Klein Goldewijk, Kees, A. Beusen, and P. Janssen. 2010. Long-Term Dynamic Modeling of Global Population and Built-Up Area in a Spatially Explicit Way, HYDE 3.1. *The Holocene* 20(4):565-573, doi: 10.1177/0959683609356587.

Klein Goldewijk, Kees, A. Beusen, M. de Vos, and G. van Drecht. 2011. The HYDE 3.1 Spatially Explicit Database of Human-Induced Land Use Change over the Past 12,000 Years. *Global Ecology and Biogeography* 20(1):73-86, doi: 10.1111/j.1466-8238.2010.00587.x.

Kolbert, Elizabeth. 2014. *The Sixth Extinction: An Unnatural History*. London: Bloomsbury Publishing.

Krause, Bernie. 2001. Loss of Natural Soundscape: Global Implications of Its Effect on Humans and Other Creatures. Speech at San Francisco World Affairs Council, January 31.

Krause, Bernie. 2002. *Wild Soundscapes: Discovering the Voice of the Natural World*. Yale University Press.

Krause, Bernie. 2008. Anatomy of the Soundscape. *Journal of the Audio Engineering Society* 56(1/2, January):73-80.

Landers, D., and A. Nahlik. 2013. *Final Ecosystem Goods and Services Classification System* (FEGS-CS). EPA/600/R-13/ORD-004914, U.S. Environmental Protection Agency, Washington, DC.

Le News. 2016. How Many Planets Would Be Needed If Everyone Lived Like the Swiss? March 23.

Leigh, G. J. 2004. *The World's Greatest Fix: A History of Nitrogen and Agriculture*. Oxford, U.K.: Oxford University Press.

Leopold, Aldo. 1949. *A Sand County Almanac: With Essays on Conservation from Round River*. New York: Oxford University Press.

Levallois, Clement. 2010. Can De-Growth be Considered a Policy Option? A Historical Note on Nicholas Georgescu-Roegen and the Club of Rome. *Ecological Economics* 69:2271-2278, doi: 10.1016/j.ecolecon.2010.06.020.

Lloyd, William Forster. 1833. *Two Lectures on the Checks to Population*. Oxford University, Oxford, U.K.

Lunde, Darrin. 2017. *The Naturalist: Theodore Roosevelt, A Lifetime of Exploration, and the Triumph of American Natural History*. New York: Broadway Books.

Maes, Joachim, N. D. Crossman, and Benjamin Burkhard. 2016. Mapping

Ecosystem Services. In *Handbook of Ecosystem Services*, ed. M. Potschin, R. Haines-Young, R. Fish, and R. K. Turner, 188-204. London: Routledge.

Maes, Joachim, Benis Egoh, Louise Willemen, Camino Liquete, Petteri Vihervaara, Jan Philipp Schägner, et al. 2012. Mapping Ecosystem Services for Policy Support and Decision Making in the European Union. Ecosystem Services 1(1):31-39, doi: 10.1002/2016GL071930.

Maes, Joachim, Anne Teller, Markus Erhard, Bruna Grizzetti, José I. Barredo, Maria Luisa Paracchini, et al. 2018. *Mapping and Assessment of Ecosystems and Their Services: An Analytical Framework for Mapping and Assessment of Ecosystem Condition.* Publications Office of the European Union, Luxembourg.

Mann, Charles C. 2005. *1491: New Revelations of the Americas Before Columbus.* New York: Alfred A. Knopf.

Mann, Charles C. 2018. Can Planet Earth Feed 10 Billion People? Humanity Has 30 years to Find Out. *The Atlantic.* March.

Mayumi, Kozo. 2001. *The Origins of Ecological Economics: The Bioeconomics of Geogescu-Roegen.* New York: Routledge.

McCord, Edward L. 2012. *The Value of Species.* New Haven: Yale University Press.

McKibben, Bill. 1989. *The End of Nature.* New York: Penguin Random House.

McKibben, Bill. 2007. *Fight Global Warming Now: The Handbook for Taking Action in Your Community.* New York: St. Martin's Press.

McMahon, G., S. M. Gregonis, S. W. Waltman, James M. Omernik, T. D. Thorson, J. A. Freeouf, et al. 2001. Developing a Spatial Framework of Common Ecological Regions for the Conterminous United States. *Environmental Management* 28(3):293-316.

Meadows, Dennis L. 1974. *Dynamics of Growth in a Finite World.* Cambridge, MA: Wright-Allen Press.

Meadows, Donella H., Dennis L. Meadows, Jørgens Randers, and William W. Behrens III. 1974. *The Limits to Growth: A Report for the Club of Rome's Project on the Predicament of Mankind*, 2nd ed. New York: Potomac Associates.

Miller, Char. 2004. *Gifford Pinchot and the Making of Modern Environmentalism (Pioneers of Conservation).* Washington, DC: Island Press.

Mononen L., A. P. Auvinen, A. L. Ahokumpu, M. Ronka, N. Aarras, H. Tolvanen, et al. 2016. National Ecosystem Service Indicators: Measures of Social-Ecological Sustainability. *Ecological Indicators* 61:27-37.

Mooney, Chris. 2018. Melting Arctic Ice Is Now Pouring 14,000 Tons of Water per Second into the Ocean, Scientists Find. *Washington Post,* December 21.

Muir, John. 1980. *Wilderness Essays.* Layton, UT: Gibbs M. Smith, Inc.

Myers, Norman, ed. 1984. *Gaia: An Atlas of Planetary Management.* London: Gaia Books.

Nash, Roderick. 1967. *Wilderness and the American Mind.* Stevens Point: University of Wisconsin Press.

Nature Needs Half. https://natureneedshalf.org.

Nelson, Richard R., and Sidney G. Winter. 1982. *An Evolutionary Theory of Economic Change*. Cambridge, MA: Belknap Press of Harvard University Press.

NOAA Headquarters. 2015. New Research Maps Areas Most Vulnerable to Ocean Acidification. *ScienceDaily*, October 13.

Nunez, Christina. 2016. The World Is Hemorrhaging Methane, and Now We Can See Where. *National Geographic*, January 13.

Olson, David M., and Eric Dinerstein. 1998. The Global 200: A Representation Approach to Conserving the Earth's Most Biologically Valuable Ecoregions. *Conservation Biology* 12:502-515.

Olson, David M., Eric Dinerstein, Eric D. Wikramanayake, Neil D. Burgess, George V. N. Powell, Emma C. Underwood, et al. 2001. Terrestrial Ecoregions of the World: A New Map of Life on Earth. *BioScience* 51(11, November):933.

Omernik, James M. 1987. Ecoregions of the Conterminous United States (map supplement, 1:7,500,000 scale). *Annals of the Association of American Geographers* 77(1):118-125.

Omernik, James M. 1995a. Ecoregions: a Spatial Framework for Environmental Management. In *Biological Assessment and Criteria, Tools for Water Resource Planning and Decision Making*, ed. W. S. Davis and T. P. Simon, 31-47. Boca Raton, FL: Lewis Publishers.

Omernik, James M. 1995b. Ecoregions: a Framework for Managing

Ecosystems. *The George Wright Forum* 12(1):35-50.

Omernik, James M. 2004. Perspectives on the Nature and Definition of Ecological Regions. *Environmental Management* 34(Supplement 1):S27-S38.

Omernik, James M., and A. L. Gallant. 1990. Defining Regions for Evaluating Environmental Resources. In *Global Natural Resource Monitoring and Assessments: Preparing for the 21st Century – Proceedings of the International Conference and Workshop* (September 24-30, 1989, Fondazione G. Cini, Venice, Italy), 2: 936-947. Bethesda, MD: American Society for Photogrammetry and Remote Sensing.

Omernik, James M., and G. E. Griffith. 2014. Ecoregions of the Conterminous United States: Evolution of a Hierarchical Spatial Framework. *Environmental Management* 54(6):1249-1266, doi: 10.1007/s00267-014-0364-1.

Oschlies, Andreas, Peter Brandt, Lothar Stramma, and Sunke Schmidtko. 2018a. Drivers and Mechanisms of Ocean Deoxygenation. *Nature GeoScience* 11(July):467-473, doi: 10.1038/s41561-018-0152-2.

Oschlies, Andreas, Peter Brandt, Lothar Stramma, and Sunke Schmidtko. 2018b. How Global Warming Is Causing Ocean Oxygen Levels to Fall. *Carbon Brief*, June 15.

Paine, Thomas. 1797. Agrarian Justice. Paris.

Papworth, S. K., J. Rist, L. Coad, and E. J. Milner-Gulland. 2009. Evidence for Shifting Baseline Syndrome in Conservation. *Conservation Letters* 2:93-100.

Pauly, Daniel. 1995. Anecdotes and the Shifting Baseline Syndrome of

Fisheries. *Trends in Ecology and Evolution.* 10(10, October 1):430.

Pauly, Daniel. 2010. The Ocean's Shifting Baseline.
https://www.ted.com/talks/daniel_pauly_the_ocean_s_shifting_baselin
e?language=en.

Pengra, Bruce. 2012. *One Planet, How Many People? A Review of Earth's Carrying Capacity.* United Nations Environmental Program. June.

Penna, Anthony, N. 2010. *The Human Footprint: A Global Environmental History.* Malden, MA: Wiley-Blackwell.

Phillips, John C., and Harold Coolidge. 1934. *The First Five Years: The American Committee for International Wildlife Protection.* Cambridge, MA: American Committee for International Wildlife Protection.

Pinchot, Gifford. 1910. *The Fight for Conservation.* New York: Doubleday, Page and Company.

Potschin, M. R. Haines-Young, R. Fish, and R. K. Turner, eds. 2016. *Handbook of Ecosystem Services.* London: Routledge.

Rees, William, and Mathis Wackernagel. 1996. *Our Ecological Footprint: Reducing Human Impact on the Earth.* Philadelphia: New Society Publishers.

Reid, Carlton. 2015. *Roads Were Not Built for Cars: How Cyclists Were the First to Push for Good Roads and Became Pioneers of Motoring.* Washington, DC: Island Press.

RESOLVE EcoRegions 2017. https://ecoregions2017.appspot.com/

ResourceWatch. www.resourcewatch.org

Ricketts, Taylor H., et al. 1999. *Terrestrial Ecoregions of North America: A Conservation Assessment*. World Wildlife Fund. Washington, DC: Island Press.

Robinson, Kim Stanley. 2018. Empty Half the Earth of Its Humans. It's the Only Way to Save the Planet. *The Guardian*, March 20.

Rockström, J., W. Steffen, K. Noone, Å. Persson, F. S. Chapin III, E. Lambin, et al. 2009. Planetary Boundaries: Exploring the Safe Operating Space for Humanity (PDF), *Ecology and Society* 14(2):32.

See also Rockström, J., 2010. Let the Environment Guide Our Development. TED Talk. www.youtube.com/watch?v=RgqtrlixYR4.

Sabin, Paul. 2013. *The Bet: Paul Ehrlich, Julian Simon, and Our Gamble over Earth's Future*. New Haven: Yale University Press.

Sachs, Jeffrey. 2006. *The End of Poverty: Economic Possibilities for Our Time*. New York: Penguin Books.

Sauer, Carl O. 1968. *Agricultural Origins and Dispersals: The Domestication of Animals and Foodstuffs*. Cambridge, MA: MIT Press.

Schafer, R. Murray. 1977. *The Tuning of the World*. New York: Random House.

Scheer, Roddy, and Doug Moss. What Causes Ocean 'Dead Zones'? *Scientific American*, n.d.

Schumpeter, Joseph. 1942. *Capitalism, Socialism and Democracy.* New York: Harper and Brothers.

Schwab, Klaus. 2016. *The Fourth Industrial Revolution.* Geneva: World Economic Forum.

Scott, James. 2011. "Four Domestications: Fire, Plants, Animals, and ... Us." Tanner Lecture on Human Values at Harvard University, Cambridge, MA.

Searchinger, Tim, Craig Hanson, Richard Waite, Sarah Harper, George Leeson, and Brian Lipinski. 2013. Achieving Replacement-Level Fertility. Washington, DC: World Resources Institute.

Smil, Vaclav. 2001. *Enriching the Earth: Fritz Haber, Carl Bosch, and the Transformation of World Food Production.* Cambridge, MA: MIT Press.

Smith, Michael B. 1998. The Value of a Tree: Public Debates of John Muir and Gifford Pinchot. *Historian* 60(4):757-778.

Smith, Neil. 2003. American Empire: Roosevelt's Geographer and the Prelude to Globalization. *California Studies in Critical Human Geography* 9(xvii):557.

Soga, Masashi, and Kevin J. Gaston. 2018. Shifting Baseline Syndrome: Causes, Consequences, and implications. *Frontiers in Ecology and the Environment* 16(4 April):222-230.

Solow, R. 1956. A Contribution to the Theory of Economic Growth. *The Quarterly Journal of Economics,* 70(1):65.

Spalding, Mark D., Helen E. Fox, Gerald R. Allen, Nick Davidson, Zach A.

Ferdaña, Max Finlayson, et al. 2007. Marine Ecoregions of the World: A Bioregionalization of Coastal and Shelf Areas. *BioScience* 57(7, 1 July): 573–583, doi: 10.1641/B570707.

Stager, Curt. 2018. The Silence of the Bugs. *New York Times*, May 26.

Steffen, Will, Katherine Richardson, Johan Rockström, Sarah E. Cornell, Ingo Fetzer, Elena M. Burnett, et al. 2015. Planetary Boundaries: Guiding Human Development on a Changing Planet. *Science* 347(6223):1259855.

Sutton, P, D. Roberts, C. Elvidge, and K. Baugh. 2001. Census from Heaven: An Estimate of the Global Human Population Using Nighttime Satellite Imagery. *International Journal of Remote Sensing* 22(16):3061-3076.

Sutton, Paul. 2003. An Empirical Environmental Sustainability Index Derived Solely from Nighttime Satellite Imagery and Ecosystem Service Values. *Population and Environment* 24(4):293-311.

Sutton, Paul, and Robert Costanza. 2002. Global Estimates of Market and Non-Market Values Derived from Nighttime Satellite Imagery, Land Use, and Ecosystem Service Valuation. *Ecological Economics* 41(3):509-527.

Sutton, Paul C., Sharolyn J. Anderson, Christopher D. Elvidge, Benjamin T. Tuttle, and Tilottama Ghosh. 2009. Paving the Planet: Impervious Surface as Proxy Measure of the Human Ecological Footprint. *Progress in Physical Geography: Earth and Environment* 33(4, August 1):510-527.

Sutton, Paul C., Sharolyn J. Anderson, Benjamin T. Tuttle, and Lauren Morse. 2011. The Real Wealth of Nations: Mapping and Monetizing the Human Ecological Footprint. *Ecological Indicators* 16(May):11-22, doi: 10.1016/j.ecolind.2011.03.008.

Taleb, Nassim Nicholas. 2010. *Black Swan: The Impact of the Highly Improbable*. New York: Random House.

Tansley, Sir Arthur George. 1935. The Use and Abuse of Vegetational Concepts and Terms. *Ecology* 16:284-307.

Tenenbaum, David. 2009. Birthplace of Ecological Restoration Celebrates 75 Years. *University of Wisconsin-Madison News*, June 17.

The Half Earth Project (www.half-earthproject.org)

Thoreau, Henry David. 1854. *Walden: or, Life in the Woods*. Boston: Ticknor and Fields.

Troy, Austin, and Matthew A. Wilson. 2006. Mapping Ecosystem Services: Practical Challenges and Opportunities in Linking GIS and Value Transfer. *Ecological Economics* 60:435-449.

Turner II, B. L., William C. Clark, Robert W. Kates, John F. Richards, Jessica T. Mathews, and William B. Meyer, eds. 1993. *The Earth as Transformed by Human Action: Global and Regional Changes in the Biosphere over the Past 300 Years Revised Edition*. Cambridge, U.K.: Cambridge University Press.

U.S. Environmental Protection Agency. 2013. *Level III and IV Ecoregions of the Continental United States* (map, 1:3,000,000 scale). EPA, National Health and Environmental Effects Research Laboratory, Corvallis, Oregon, https://www.epa.gov/eco-research/level-iii-and-iv-ecoregions-continental-united-states.

U.S. Environmental Protection Agency. 2015. National Ecosystem Services

Classification System (NESCS): Framework Design and Policy Application. EPA-800-R-15-002. EPA, Washington, DC.

United Nations Environment Program, World Conservation Monitoring Centre and International Union for Conservation of Nature. 2016. *Protected Planet Report 2016*. Cambridge, U.K. and Gland, Switzerland: UNEP-WCMC and IUCN.

United Nations Population Fund. 2016. *Universal Access to Reproductive Health: Progress and Challenges*. New York: UNPF.

United Nations. 2018. World Urbanization Prospects 2018. https://www.un.org/development/desa/publications/2018-revision-of-world-urbanization-prospects.html.

United Nations. World Database on Protected Areas. www.protectedplanet.net.

Van Andel, Tjeerd H., Eberhard Zangger, and Anne Demitrack. 2013. Land Use and Soil Erosion in Prehistoric and Historical Greece. *Journal of Field Archaeology* 17(4):379-396.

Vyawahare, Malavika. 2018. Can the Planet Support 11 Billion People? By the End of This Century, That Many People May Be Inhabiting This Planet, According to the Latest U.N. Projections. *Scientific American*, August 12. https://www.scientificamerican.com/article/can-the-planet-support-11-billion-people/.

Williams, Michael. 2006. *Deforesting the Earth: From Prehistory to Global Crisis, an Abridgment*. Chicago: The University of Chicago Press.

Wilson, E. O. 2016. *Half-Earth: Our Planet's Fight for Life*. New York: W.W. Norton and Company.

Winchester, Simon. 2001. *The Map That Changed the World: William Smith and the Birth of Modern Geology*. New York: HarperCollins.

Winner, Langdon. 1978. *Autonomous Technology: Technics-out-of-Control as a Theme in Political Thought*. Cambridge, MA: MIT Press.

World Data Lab (http://www.worlddata.io/)

World Wildlife Fund. 2000. Freshwater Ecoregions of North America: A Conservation Assessment. WWF Ecoregion Assessments. Washington, DC: Island Press.

World Wildlife Fund. Biennial. *The Living Planet Report*.

Wulf, Andrea. 2015. *The Invention of Nature: Alexander von Humboldt's New World*. New York: Alfred A. Knopf.

Zeder, Melinda A. 2008. Domestication and Early Agriculture in the Mediterranean Basin: Origins, Diffusion, and Impact. *Proceedings of the National Academy of Sciences of the United States of America* 105(33):11597-11604.

Index